Test Yourself

Introduction to Calculus

Joan Van Glabek, Ph.D.
Department of Mathematics
Edison Community College
Naples, FL

Contributing Editors

Mark N. Weinfeld, M.S.
President
MATHWORKS
New York, NY

Carl E. Langenhop, Ph.D.
Emeritus Professor of Mathematics
Southern Illinois University—Carbondale
Carbondale, IL

Douglas G. Smith
Arthur P. Schalick High School
Elmer, NJ

NTC LearningWorks
a division of NTC Publishing Group
Lincolnwood, Illinois

Library of Congress Cataloging-in-Publication Data
is available from the Library of Congress.

A *Test Yourself Books, Inc.* Project

Contents

Preface

These test questions and answers have been written to help you more fully understand the material in a first-semester calculus course. Working through these test questions prior to a test will help you pinpoint areas that require further study. Each answer is followed by a parenthetical topic description that corresponds to a section in your calculus textbook. Return to your textbook as necessary to reread and review areas in which you feel unsure about your work. See your professor for additional help, and/or get together with a study group to complete your preparation for each test. Work some problems from this test book every day. This will allow you time to prepare properly for your exams. Cramming in calculus is *not* advisable.

Good luck with your study of calculus!

Joan Van Glabek, Ph.D.

How to Use this Book

This "Test Yourself" book is part of a unique series designed to help you improve your test scores on almost any type of examination you will face. Too often, you will study for a test—quiz, midterm, or final—and come away with a score that is lower than anticipated. Why? Because there is no way for you to really know how much you understand a topic until you've taken a test. The *purpose* of the test, after all, is to test your complete understanding of the material.

The "Test Yourself" series offers you a way to improve your scores and to actually test your knowledge at the time you use this book. Consider each chapter a diagnostic pretest in a specific topic. Answer the questions, check your answers, and then give yourself a grade. Then, and only then, will you know where your strengths and, more importantly, weaknesses are. Once these areas are identified, you can strategically focus your study on those topics that need additional work.

Each book in this series presents a specific subject in an organized manner, and although each "Test Yourself" chapter may not correspond to exactly the same chapter in your textbook, you should have little difficulty in locating the specific topic you are studying. Written by educators in the field, each book is designed to correspond as much as possible to the leading textbooks. This means that you can feel confident in using this book that regardless of your textbook, professor, or school, you will be much better prepared for anything you will encounter on your test.

Each chapter has four parts

Brief Yourself. All chapters contain a brief overview of the topic that is intended to give you a more thorough understanding of the material with which you need to be familiar. Sometimes this information is presented at the beginning of the chapter, and sometimes it flows throughout the chapter, to review your understanding of various *units* within the chapter.

Test Yourself. Each chapter covers a specific topic corresponding to one that you will find in your textbook. Answer the questions, either on a separate page or directly in the book, if there is room.

Check Yourself. Check your answers. Every question is fully answered and explained. These answers will be the key to your increased understanding. If you answered the question incorrectly, read the explanations to *learn* and *understand* the material. You will note that at the end of every answer you will be referred to a specific subtopic within that chapter, so you can focus your studying and prepare more efficiently.

Grade Yourself. At the end of each chapter is a self-diagnostic key. By indicating on this form the numbers of these questions you answered incorrectly, you will have a clear picture of your weak areas.

There are no secrets to test success. Only good preparation can guarantee higher grades. By utilizing this "Test Yourself" book, you will have a better chance of improving your scores and understanding the subject more fully.

Algebra Review

Brief Yourself

This chapter contains a review of the algebra topics that will be used during the remainder of the book. In this chapter you will find questions about the real number system, equations and inequalities, and introductory graphing topics involving lines and circles.

The real number system contains two types (or sets) of numbers—rational numbers and irrational numbers. A rational number can be written as a ratio of two integers, such as 1/2, –3/4, 4/3, 0.3333... (which can be written as 1/3), etc. Numbers such as $\sqrt{3}$, π, e are irrational numbers. Irrational numbers are nonrepeating, nonterminating decimals.

To solve first-degree equations (that is, equations in which the variable is raised to the first power), isolate the variable on one side of the equation by adding (or subtracting) the same quantity to (from) both sides of the equation. Then divide both sides of the equation by the coefficient of the variable. To solve a first-degree inequality, follow the same basic rules as for a first-degree equation, remembering that multiplying or dividing by a negative number causes the direction of the inequality to change (for example, from < to >).

Some of the basic formulas needed for working with coordinate geometry include the following (*all* of these formulas should be committed to memory):

Distance between points $P_1(x_1, y_1)$ and $P_2(x_2, y_2) = \sqrt{(x_2 - x_1)^2 + (y_2 - y_1)^2}$

Midpoint of the line segment joining $P_1(x_1, y_1)$ and $P_2(x_2, y_2) = \left(\dfrac{x_1 + x_2}{2}, \dfrac{y_1 + y_2}{2} \right)$ (Note that to find the midpoint you find the average of the x coordinates, and the average of the y coordinates.)

The slope of a line through $P_1(x_1, y_1)$ and $P_2(x_2, y_2)$: $m = \dfrac{y_2 - y_1}{x_2 - x_1} = \dfrac{\text{change in y}}{\text{change in x}}$

Equation in standard form of a circle with center (x_1, y_1) and radius r: $(x - x_1)^2 + (y - y_1)^2 = r^2$

Equations of Lines:

General form: $ax + by + c = 0$ Standard form: $ax + by = c$

Point-slope form: $y - y_1 = m(x - x_1)$ Slope-intercept form: $y = mx + b$

Equations of Lines (*cont.*):

Equation of a horizontal line through (x_1, y_1) : $y = y_1$

If two lines are parallel, they have equal slopes. If two lines are perpendicular and neither has slope 0, they have slopes that are negative reciprocals of each other (for example, 1/2 and –2).

Remember that graphing techniques are often short-cuts for the longer method of plotting points. However, if you cannot recall a specific technique, plot as many points as time allows to help you graph an equation.

Test Yourself

1. Which of the following are rational numbers?
$$-5, -1.4, -\pi, 0, \frac{1}{2}, \sqrt{4}, \frac{9}{5}$$

2. Classify each number as rational or irrational.

   a) 1.333...

   b) $\sqrt{3}$

   c) π

Use interval notation to describe each set.

3. $\{x | x > -3\}$

4. $\{x | -2 < x \le 4\}$

5. $\{x | x \le 0\}$

Graph each set on a number line.

6. $[-2, 3]$

7. $(1, 4)$

8. $(-\infty, 2]$

9. $(-1, \infty)$

Solve each inequality.

10. $4x - 8 > 12$

11. $-2x + 6 \le 2$

12. $-\frac{1}{2}x - 4 \ge 3$

13. $4 < 3x - 2 < 10$

14. $-3 \le 2x - 5 \le 8$

15. $x + 2 < 3x + 5 < 6$

Solve.

16. $|x - 6| = 8$

17. $|2x - 3| = -6$

18. $|4 - 5x| = 2$

19. $|3x + 4| = 0$

Solve. Write each solution in interval notation.

20. $|1 + 4x| < 7$

21. $|3x - 5| \ge 4$

22. $|2 - 5x| \le -3$

23. $|x + 4| > -6$

24. $\left|4 - \frac{1}{2}x\right| \le 10$

Solve each inequality. Write each solution in interval notation.

25. $x^2 + 2x > 8$

26. $2x^2 \le 5x + 12$

27. $2x^2 + 4x < 7$

Find the exact distance between the given points.

28. $(6, -5)$ and $(-3, -9)$

29. $(3\sqrt{2}, 4)$ and $(-2\sqrt{2}, -1)$

30. $\left(\frac{1}{2}, -\frac{1}{2}\right)$ and $\left(-\frac{2}{5}, 4\right)$

31. Find the distance between the given points. Round your answer to the nearest hundredth.
    $(-2.5, -7.1)$ and $(1.8, -2.3)$

32. Find the midpoint of the line segment joining the given points. $(4, -6)$ and $(-3, -8)$

33. Find the midpoint of the line segment joining the given points. $(2\sqrt{5}, \sqrt{3})$ and $(8\sqrt{5}, -4\sqrt{3})$

Find the equation of each circle. Write each answer in standard form.

34. Center at the origin and radius 9

35. Center at $(0, 4)$ and radius 5

36. Center $(-3, -5)$ and radius 16

37. Endpoints of a diameter: $(3, 7)$ and $(-5, 3)$

Find the center and radius of each circle.

38. $x^2 + y^2 + 4x - 6y - 9 = 0$

39. $x^2 + y^2 = 25$

40. $x^2 + y^2 - x + 3y - 5 = 0$

41. $2x^2 + 2y^2 - 6x + 3y - 10 = 0$

42. Find the slope of the line containing the points $(-6, -2)$ and $(-8, 4)$.

43. Find the slope of the line $4x - 5y = 12$.

44. Find the slope of a line parallel to $x + 3y = 6$.

45. Find the slope of a line perpendicular to $x + 3y = 6$.

46. Find the equation of the line with slope $-\frac{2}{3}$ containing the point $(-3, 5)$.

47. Find the equation of the horizontal line containing the point $(-1, 3)$.

48. Find the equation of the vertical line containing the point $(-1, 3)$.

49. Find the equation of the line containing $(-2, 6)$ and $(-2, -1)$.

50. Find the equation of the line with zero slope containing the point $(4, -1)$.

51. Find the equation of the line with undefined slope that contains the point $(0, 3)$.

52. Find the equation of the line parallel to $2x - 5y = 8$ and containing the point $(-1, 4)$. Write your answer in slope-intercept form.

53. Find the equation of the line perpendicular to $3x + y = 4$ and containing the point $(2, -3)$. Write your answer in slope-intercept form.

54. Find the equations of the lines tangent to $x^2 + y^2 = 4$ at $x = 1$.

55. Find the equation of the line parallel to $x = -2$ and containing the point $(4, 6)$.

56. Find the equation of the line perpendicular to $y + 4 = 0$ and containing the point $(-3, 1)$.

Graph each equation.

57. $y = x^2 + 2$

58. $y = |x - 1|$

59. $y = \sqrt{x} - 3$

60. $y = \frac{2}{3}x - 4$

61. $2x - 3y = 6$

62. $x^2 + y^2 = 16$

63. $(x + 1)^2 + (y - 2)^2 = 9$

64. $y = (x - 1)^2 + 3$

65. $y = 2$

66. $x^2 + y^2 + 4x - 2y = 10$

67. Determine whether the graph of $x^2 + y^2 = 1$ is symmetric with respect to the x-axis, y-axis, origin, or none of these.

68. Determine whether the graph of $y = \frac{1}{2}x + 4$ is symmetric with respect to the x-axis, y-axis, origin, or none of these.

69. Determine whether the graph of $y = x^2 - 3$ is symmetric with respect to the x-axis, y-axis, origin, or none of these.

70. Determine whether the graph of $y = x^5 + x^3$ is symmetric with respect to the x-axis, y-axis, origin, or none of these.

Determine whether the given equation represents the graph of a line, a parabola, or a circle.

71. $y = -x^2 + 4$

72. $3x + y = 6$

73. $x^2 + y^2 - 4x + 6y = 8$

74. $x = 3$

75. $2y - 5 = 7$

Check Yourself

1. A rational number can be written as a ratio of two integers. Thus, $-5, -1.4 = -\frac{14}{10}, 0 = \frac{0}{a}$ for $a \neq 0, \frac{1}{2}$ and $\frac{9}{5}$ are rational numbers. Since $\sqrt{4} = 2 = \frac{2}{1}$, $\sqrt{4}$ is also a rational number. **(Real Numbers)**

2. a) Rational. Nonterminating, repeating decimals can be written as rational numbers. $1.333... = 1\frac{1}{3} = \frac{4}{3}$. **(Real Numbers)**

    b) Irrational. Note that only the square root of a perfect square (such as $\sqrt{4}$) would be rational. **(Real Numbers)**

    c) Irrational. Although π is often *approximated* by $\frac{22}{7}$ or 3.14, π is a nonrepeating, nonterminating decimal and hence is an irrational number. **(Real Numbers)**

3. $(-3, \infty)$ Endpoints excluded from the interval are always enclosed in parentheses. Note that ∞ is always followed by a parenthesis. **(Interval Notation)**

4. $(-2, 4]$ End points included in the interval are enclosed in brackets. **(Interval Notation)**

5. $(-\infty, 0]$ $-\infty$ is always preceded by a parenthesis, never a bracket. **(Interval Notation)**

6.
```
     [————————]
    -2         3
```
(Interval Notation)

7.
```
     (————————)
     1         4
```
(Interval Notation)

8.
```
   ◄————————]
            2
```
(Interval Notation)

9.   **(Interval Notation)**

10. $4x - 8 > 12$ Given

    $4x > 20$ Add 8 to both sides of the inequality

    $x > 5$ Divide both sides by 4; do not change the direction of the inequality symbol

    $(5, \infty)$ **(Solving First-degree and Compound Inequalities)**

11. $-2x + 6 \leq 2$ Given

    $-2x \leq -4$ Subtract 6 from both sides of the inequality

    $x \geq 2$ Divide both sides by –2. Notice the inequality symbol changes from $\leq$ to $\geq$

    $[2, \infty)$ **(Solving First-degree and Compound Inequalities)**

12. $-\frac{1}{2}x - 4 \geq 3$ Isolate x by first adding 4 to both sides of the inequality

    $-\frac{1}{2}x \geq 7$ Multiply both sides of the inequality by $-\frac{2}{1}$. Remember to reverse the inequality symbol

    $x \leq -14$ Notice that the $\geq$ has now become $\leq$

    $(-\infty, -14]$ Write the solution in interval notation, using a bracket on –14 to show that it is included in the solution. **(Solving First-degree and Compound Inequalities)**

13. $4 < 3x - 2 < 10$ Given

    $6 < 3x < 12$ Isolate x-term by adding 2 to all three parts

    $2 < x < 4$ Divide each part by 3

    $(2, 4)$ Write the solution in interval notation, using parentheses to show that neither endpoint is included in the solution. **(Solving First-degree and Compound Inequalities)**

14. $-3 \leq 2x - 5 \leq 8$ Given

    $2 \leq 2x \leq 13$ Add 5 to all three parts

    $1 \leq x \leq \frac{13}{2}$ Divide each part by 2

    $[1, \frac{13}{2}]$ Write the solution in interval notation, using brackets to show that both endpoints are included in the answer. **(Solving First-degree and Compound Inequalities)**

15. Since x appears in more than one part of this compound inequality, split the inequality into two parts connected by the word "and."

    $x + 2 < 3x + 5 < 6$ Given

    $x + 2 < 3x + 5$ and $3x + 5 < 6$ Split into two inequalities, joined by "and"

    $-2x < 3$ and $3x < 1$ Isolate the x-term

$x > -\dfrac{3}{2}$ and $x < \dfrac{1}{3}$ Solve for x

Because "and" represents an intersection, the solution is $\left(-\dfrac{3}{2}, \dfrac{1}{3} \right)$. **(Solving First-degree and Compound Inequalities)**

16. $|x - 6| = 8$ Given

    $x - 6 = 8$ or $x - 6 = -8$ Set up two equations using the fact that $|x| = a$ if and only if $x = a$ or $x = -a$, $(a \geq 0)$

      $x = 14$ or $x = -2$ Solve each equation for x

The solution set is $\{-2, 14\}$. Note that the solution cannot be written as an interval because the solution is only two points, -2 and 14. **(Absolute Value)**

17. $|2x - 3| = -6$ has no solutions, which can be written as $\varnothing$, since no value of x can make an absolute value equal a negative number. **(Absolute Value)**

18. $|4 - 5x| = 2$ Given

    $4 - 5x = 2$ or $4 - 5x = -2$ Set up two equations

     $-5x = -2$ $-5x = -6$ Solve each equation for x

      $x = 2/5$ $x = 6/5$ **(Absolute Value)**

19. $|3x + 4| = 0$ Given

    $3x + 4 = 0$ Because $\pm 0 = 0$, you only need to solve one equation

     $x = -4/3$ **(Absolute Value)**

20. $|1 + 4x| < 7$ Given

    $-7 < 1 + 4x < 7$ Rewrite without absolute value bars using $|x| < a$ if and only if $-a < x < a$, $a > 0$

    $-8 < 4x < 6$ Subtract 1 from each part

    $-2 < x < \dfrac{3}{2}$ Divide each part by 4

The answer in interval notation is $\left(-2, \dfrac{3}{2} \right)$ **(Absolute Value)**

21. $|3x - 5| \geq 4$ Given

    $3x - 5 \geq 4$ or $3x - 5 \leq -4$ Rewrite without absolute value bars using $|x| \geq a$ if and only if $x \geq a$ or $x \leq -a$

     $3x \geq 9$ or $3x \leq 1$ Solve each inequality

      $x \geq 3$ or $x \leq \dfrac{1}{3}$ Divide by 3

The answer in interval notation is $\left(-\infty, \dfrac{1}{3}\right] \cup [3, \infty)$. **(Absolute Value)**

22. $|2 - 5x| \le -3$ has no solutions (which can be written $\varnothing$) since no value of x can make an absolute value less than or equal to a negative number. **(Absolute Value)**

23. $|x + 4| > -6$ has solution set of all real numbers (written $(-\infty, \infty)$) since every value of x will make this absolute value greater than a negative number. **(Absolute Value)**

24. $\left|4 - \dfrac{1}{2}x\right| \le 10$ Given

$-10 \le 4 - \dfrac{1}{2}x \le 10$ Rewrite without absolute value bars using $|x| \le a$ if and only if $-a \le x \le a$, $a > 0$

$-14 \le -\dfrac{1}{2}x \le 6$ Add -4 to each part

$28 \ge x \ge -12$ Multiply by -2. Reverse the inequality symbols when multiplying by a negative number

The solution is $[-12, 28]$. Be careful to write the smaller number first in interval notation. Use brackets to show that the endpoints are included in the solution. **(Absolute Value)**.

25. $x^2 + 2x > 8$ Given

$x^2 + 2x - 8 > 0$ Get 0 on the right by subtracting 8 from both sides

$(x + 4)(x - 2) > 0$ Factor

Set each factor equal to 0 and solve to find the split points. This gives split points of -4 and 2. Draw a number line, mark the split points, and find the sign of each factor in each region formed by the split points:

Because the original inequality required values greater than 0, use the positive regions for the answer: $(-\infty, -4) \cup (2, \infty)$.

You can also solve the problem by using a graphing calculator to graph $y = x^2 + 2x - 8$. Note that when the graph is above the x-axis, $y > 0$, which means $x^2 + 2x - 8 > 0$. This will occur when $x < -4$ or when $x > 2$. **(Quadratic Inequalities)**

26. $2x^2 \le 5x + 12$ Given

$2x^2 - 5x - 12 \le 0$ Subtract $5x + 12$ from both sides

$(2x + 3)(x - 4) \le 0$ Factor

Set each factor equal to 0 and solve to find the split points. This gives split points of $-3/2$ and 4. Draw a number line, mark the split points, and find the sign of each factor in each region:

$$2x + 3 \qquad - \qquad + \qquad +$$

$$x - 4 \qquad - \qquad - \qquad +$$

$$(2x + 3)\ (x - 4) \qquad \oplus \qquad \ominus \qquad \oplus$$

Because the original inequality required values less than or equal to 0, use the negative region for the answer: $[-\dfrac{3}{2}, 4]$. Note that brackets (rather than parentheses) are used to show that the endpoints are included in the answer.

You can also use a graphing calculator to graph $y = 2x^2 - 5x - 12$. Since we need $y \le 0$, look for values where the graph falls below the x-axis. These will occur when $-\dfrac{3}{2} \le x \le 4$. **(Quadratic Inequalities)**

27. $2x^2 + 4x < 7$ Given

$2x^2 + 4x - 7 < 0$ Subtract 7 from both sides

These split points cannot be found by factoring. Using the quadratic formula, we have:

$$x = \dfrac{-4 \pm \sqrt{16 - 4\,(2)\,(-7)}}{2\,(2)} \qquad a = 2, b = 4, c = -7$$

$$= \dfrac{-4 \pm \sqrt{72}}{4} \qquad \text{Simplify the radicand}$$

$$= \dfrac{-4 \pm 6\sqrt{2}}{4} \qquad (\text{Since } \sqrt{72} = \sqrt{36} \cdot \sqrt{2} = 6\sqrt{2})$$

$$= \dfrac{-2 \pm 3\sqrt{2}}{2}$$

Use your calculator to approximate these values as -3.12 and 1.12. Test around these split points:

When $x = -4$, $2x^2 + 4x - 7 = 2\,(-4)^2 + 4\,(-4) - 7 < 0$, False

When $x = 0$, $2x^2 + 4x - 7 = 2\,(0)^2 + 4\,(0) - 7 < 0$, True

When $x = 2$, $2x^2 + 4x - 7 = 2\,(2)^2 + 4\,(2) - 7 < 0$, False

The solution is $\left[\dfrac{-2 - 3\sqrt{2}}{2}, \dfrac{-2 + 3\sqrt{2}}{2}\right]$ **(Quadratic Inequalities)**

28. Use the distance formula with $x_1 = -3$, $x_2 = 6$, $y_1 = -9$, $y_2 = -5$

$$d = \sqrt{(6 + (-3))^2 + (-5 - (-9))^2} = \sqrt{(9)^2 + (4)^2} = \sqrt{97}.$$ With $x_1 = 6$, $x_2 = -3$, $y_1 = -5$, $y_2 = -9$, one gets the same value.

The word "exact" in the directions means that you should not approximate the distance with a decimal. The answer is $\sqrt{97}$. **(The Distance and Midpoint Formulas)**

29. $d = \sqrt{\left(3\sqrt{2}-(-2\sqrt{2})\right)^2 + \left(4-(-1)\right)^2} = \sqrt{\left(5\sqrt{2}\right)^2 + 5^2} = \sqrt{50+25} = \sqrt{75} = 5\sqrt{3}$ **(The Distance and Midpoint Formulas)**

30. $d = \sqrt{\left(-\dfrac{2}{5}-\dfrac{1}{2}\right)^2 + \left(4-\left(-\dfrac{1}{2}\right)\right)^2}$ Use the distance formula

$= \sqrt{\left(-\dfrac{4}{10}-\dfrac{5}{10}\right)^2 + \left(\dfrac{8}{2}+\dfrac{1}{2}\right)^2}$ Get common denominators

$= \sqrt{\left(-\dfrac{9}{10}\right)^2 + \left(\dfrac{9}{2}\right)^2}$ Simplify inside the parentheses

$= \sqrt{\dfrac{81}{100}+\dfrac{81}{4}}$ Square each fraction

$= \sqrt{\dfrac{81}{100}+\dfrac{2025}{100}}$ Get a common denominator

$= \sqrt{\dfrac{2106}{100}}$ Add fractions

$= \dfrac{\sqrt{81 \cdot 26}}{10}$ Simplify the radical

$= \dfrac{9\sqrt{26}}{10}$ **(The Distance and Midpoint Formulas)**

31. $d = \sqrt{\left(1.8-(-2.5)\right)^2 + \left(-2.3-(-7.1)\right)^2}$ Use the distance formula

$= \sqrt{4.3^2 + 4.8^2}$ Simplify inside each set of parentheses

$= \sqrt{41.53} \approx 6.444$ Use a calculator to get an approximate square root

Rounding this answer to the nearest hundredth gives 6.44. Note that this answer is an approximation, *not* an exact answer. **(The Distance and Midpoint Formulas)**

32. $M = \left(\dfrac{4+(-3)}{2}, \dfrac{-6+(-8)}{2}\right) = \left(\dfrac{1}{2}, -7\right)$ **(The Distance and Midpoint Formulas)**

33. $M = \left(\dfrac{2\sqrt{5}+8\sqrt{5}}{2}, \dfrac{\sqrt{3}-4\sqrt{3}}{2}\right) = \left(5\sqrt{5}, \dfrac{-3\sqrt{3}}{2}\right)$ **(The Distance and Midpoint Formulas)**

34. A center at the origin means $(x_1, y_1) = (0, 0)$.

$(x-0)^2 + (y-0)^2 = (9)^2$ Use $(x-x_1)^2 + (y-y_1)^2 = r^2$

$x^2 + y^2 = 81$ Simplify **(Finding the Equation of a Circle)**

35. Here, $(x_1, y_1) = (0, 4)$ and $r = 5$. Substituting into $(x - x_1)^2 + (y - y_1)^2 = r^2$ gives:

$$(x - 0)^2 + (y - 4)^2 = 5^2$$

$$x^2 + (y - 4)^2 = 25 \quad \textbf{(Finding the Equation of a Circle)}$$

36. $(x - (-3))^2 + (y - (-5))^2 = 16^2$ Use $x_1 = -3$, $y_1 = -5$, $r = 16$

$$(x + 3)^2 + (y + 5)^2 = 256 \quad \text{Simplify inside the parentheses and square 16}$$

(Finding the Equation of a Circle)

37. Given the endpoints of a diameter, find the center of the circle by finding the midpoint of the diameter:
$$M = \left(\frac{3 + (-5)}{2}, \frac{7 + 3}{2} \right) = (-1, 5)$$

Find the radius by finding the distance from the center $(-1, 5)$ to one endpoint of the diameter $(3, 7)$:

$$d = \sqrt{(-1 - 3)^2 + (5 - 7)^2} = \sqrt{16 + 4} = \sqrt{20}$$

The equation of the circle is:

$$(x - (-1))^2 + (y - 5)^2 = (\sqrt{20})^2 \quad \text{Use } x_1 = -1, \, y_1 = 5, \, r = \sqrt{20}$$

$$(x + 1)^2 + (y - 5)^2 = 20 \quad \textbf{(Finding the Equation of a Circle)}$$

38. Group x^2 and x terms, y^2 and y terms. Move the constant to the right side of the equation by adding 9 to both sides of the equation. Complete the square:

$$x^2 + 4x + 4 + y^2 - 6y + 9 = 9 + 4 + 9 \quad \text{To complete the squares, add the numbers } \left(\frac{1}{2} \cdot 4 \right)^2 = 2^2 = 4 \text{ and}$$

$$\left(\frac{1}{2} \cdot (-6) \right)^2 = (-3)^2 = 9$$

$$(x + 2)^2 + (y - 3)^2 = 22 \quad \text{Factor}$$

Therefore, the center is $(-2, 3)$ and the radius is $\sqrt{22}$. **(Finding the Center and Radius of a Circle)**

39. $x^2 + y^2 = 25$ has center $(0, 0)$ and radius $\sqrt{25} = 5$. **(Finding the Center and Radius of a Circle)**

40. Use the process of completing the square to write the equation in standard form:

$$x^2 - x + y^2 + 3y = 5 \quad \text{Group the } x \text{ terms, group the } y \text{ terms, add 5 to both sides of the equation}$$

$$x^2 - x + \frac{1}{4} + y^2 + 3y + \frac{9}{4} = 5 + \frac{1}{4} + \frac{9}{4} \quad \text{To complete the squares, add } \left(\frac{1}{2} \cdot (-1) \right)^2 = \frac{1}{4} \text{ and } \left(\frac{1}{2} \cdot 3 \right)^2 = \frac{9}{4}$$

$$\left(x - \frac{1}{2} \right)^2 + \left(y + \frac{3}{2} \right)^2 = \frac{20}{4} + \frac{1}{4} + \frac{9}{4} \quad \text{Factor each of the perfect square trinomials}$$

Use a common denominator to add the numbers on the right side of the equation.

$$\left(x - \frac{1}{2}\right)^2 + \left(y + \frac{3}{2}\right)^2 = \frac{30}{4} \quad \text{Add fractions}$$

The center is $\left(\frac{1}{2}, -\frac{3}{2}\right)$ and the radius is $\sqrt{\frac{30}{4}} = \frac{\sqrt{30}}{2}$. **(Finding the Center and Radius of a Circle)**

41. Make the coefficients of the squared terms 1 to complete the square. Begin by dividing both sides of the equation (and thus each term) by 2:

$$2x^2 + 2y^2 - 6x + 3y - 10 = 0 \quad \text{Given}$$

$$x^2 + y^2 - 3x + \frac{3}{2}y - 5 = 0 \quad \text{Divide each term by 2}$$

$$x^2 - 3x + \frac{9}{4} + y^2 + \frac{3}{2}y + \frac{9}{16} = 5 + \frac{9}{4} + \frac{9}{16} \quad \text{Group } x \text{ terms, group } y \text{ terms, and complete the square on each}$$

$$\left(x - \frac{3}{2}\right)^2 + \left(y + \frac{3}{4}\right)^2 = \frac{125}{16} \quad \text{Factor, add fractions}$$

The center is $\left(\frac{3}{2}, -\frac{3}{4}\right)$ and $r = \sqrt{\frac{125}{16}} = \frac{5}{4}\sqrt{5}$ **(Finding the Center and Radius of a Circle)**

42. $m = \dfrac{y_2 - y_1}{x_2 - x_1} = \dfrac{4 - (-2)}{-8 - (-6)} = \dfrac{6}{-2} = -3$ **(Slope)**

43. Write the equation in slope-intercept form:

$$4x - 5y = 12 \quad \text{Given}$$

$$-5y = -4x + 12 \quad \text{Add } -4x \text{ to both sides of the equation}$$

$$y = \frac{4}{5}x - \frac{12}{5} \quad \text{Divide each term by } -5$$

The coefficient of x is the slope, so, $m = \dfrac{4}{5}$ **(Slope)**

44. Write $x + 3y = 6$ in $y = mx + b$ form:

$$3y = -x + 6 \quad \text{Subtract } x \text{ from both sides of the equation}$$

$$y = -\frac{1}{3}x + 2 \quad \text{Divide each term by 3}$$

This line has slope $-\dfrac{1}{3}$. Any line parallel to it has the same slope, $-\dfrac{1}{3}$. **(Slope)**

45. The given line has slope $-\dfrac{1}{3}$ (see #44). Any line perpendicular to it has slope equal to the negative reciprocal of $-\dfrac{1}{3}$, which is 3. **(Slope)**

46. Use the point-slope formula: $y - y_1 = m\ (x - x_1)$:

$$y - 5 = -\frac{2}{3}(x - (-3))$$

$$y - 5 = -\frac{2}{3}(x + 3) \quad \text{Simplify inside the parentheses}$$

On a test you may be required to write this answer in any of the following forms:

$3y - 15 = -2x - 6$ Multiply both sides of the equation by 3 to eliminate the fraction.

$2x + 3y = 9$ Add $2x$ to both sides; add 15 to both sides (standard form)

$2x + 3y - 9 = 0$ Add –9 to both sides (general form)

$3y = -2x + 9$ Isolate y to write the equation in slope-intercept form

$$y = -\frac{2}{3}x + 3 \quad \text{(slope-intercept form)} \quad \textbf{(Writing Equations of Lines)}$$

47. $y = 3$ **(Writing Equations of Lines)**

48. $x = -1$ **(Writing Equations of Lines)**

49. Note that both points contain the same x-coordinate. Therefore, this is a vertical line with equation $x = -2$.
    (Writing Equations of Lines)

50. A line with $m = 0$ is a horizontal line. Therefore, the equation is $y = -1$. **(Writing Equations of Lines)**

51. A line with undefined slope is a vertical line. Therefore, the equation is $x = 0$. **(Writing Equations of Lines)**

52. To be parallel to $2x - 5y = 8$, the line must have the same slope. To find the slope of $2x - 5y = 8$, first
    write the equation in $y = mx + b$ form:

$2x - 5y = 8$ Given

$-5y = -2x + 8$ Add $-2x$ to both sides

$$y = \frac{2}{5}x - \frac{8}{5} \quad \text{Divide each term by } -5$$

The coefficient of x is the slope, so, the required slope is 2/5. Now use the point-slope formula to find the
    equation of the line using $(x_1, y_1) = (-1, 4)$ and $m = 2/5$:

$$y - 4 = \frac{2}{5}(x - (-1))$$

$$y - 4 = \frac{2}{5}x + \frac{2}{5} \quad \text{Distribute the } \frac{2}{5}$$

$$y = \frac{2}{5}x + \frac{2}{5} + \frac{20}{5} \quad \text{Add 4 to both sides of the equation. Use a common denominator to add the fractions}$$

$$y = \frac{2}{5}x + \frac{22}{5} \quad \text{The answer is now in slope-intercept form } \textbf{(Writing Equations of Lines)}$$

53. To be perpendicular to $3x + y = 4$, the required line must have slope equal to the negative reciprocal of the slope of $3x + y = 4$. Find the slope of $3x + y = 4$ by writing it in $y = mx + b$ form: $y = -3x + 4$. Any line perpendicular to this line must have slope $-(1/-3)$ or $1/3$. Now use the point-slope formula to find the equation of the line using $(x_1, y_1) = (2, -3)$ and $m = 1/3$:

$y - (-3) = \dfrac{1}{3}(x - 2)$ Use the point-slope formula

$y + 3 = \dfrac{1}{3}x - \dfrac{2}{3}$ Simplify each side

$y = \dfrac{1}{3}x - \dfrac{2}{3} - \dfrac{9}{3}$ Add $-3 = -9/3$ to both sides of the equation

$y = \dfrac{1}{3}x - \dfrac{11}{3}$ **(Writing Equations of Lines)**

54. When $x = 1$, $1^2 + y^2 = 4$, so $y^2 = 3$ or $y = \pm\sqrt{3}$. Therefore, there are two points on the circle where $x = 1$, $(1, \sqrt{3})$ and $(1, -\sqrt{3})$. To find the slope of the tangents at these points, use the geometry theorem that says that a tangent to a circle is perpendicular to the radius at the point of tangency. We can find the slope of the radius using $(0, 0)$, the center of the circle, and the point of tangency. Thus,

$m_1 = \dfrac{\sqrt{3} - 0}{1 - 0} = \sqrt{3}$ and $m_2 = \dfrac{-\sqrt{3} - 0}{1 - 0} = -\sqrt{3}$. Since the tangent must be perpendicular to the radius, we

will find the equation of the tangent containing the point $(1, \sqrt{3})$ with slope $-\dfrac{1}{\sqrt{3}}$ for one of the tangents:

$y - \sqrt{3} = \left(-\dfrac{1}{\sqrt{3}}\right)(x - 1)$ Use the point-slope formula

$y - \sqrt{3} = \left(\dfrac{-1}{\sqrt{3}}\right)x + \dfrac{1}{\sqrt{3}}$ Distribute

$y = \left(-\dfrac{1}{\sqrt{3}}\right)x + \left(\dfrac{1}{\sqrt{3}} + \sqrt{3}\right)$ Add $\sqrt{3}$ to both sides of the equation

The other tangent contains the point $(1, -\sqrt{3})$ and has slope $\dfrac{1}{\sqrt{3}}$:

$y - (-\sqrt{3}) = \dfrac{1}{\sqrt{3}}(x - 1)$ or $y = \dfrac{1}{\sqrt{3}}x - \dfrac{1}{\sqrt{3}} - \sqrt{3}$. The graph is shown on page 14. **(Writing Equations of Lines)**

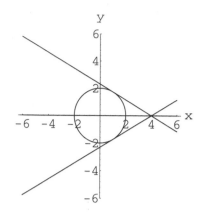

55. The line $x = -2$ is a vertical line, and any line parallel to it must also be a vertical line. Thus the required equation is $x = 4$. (**Writing Equations of Lines**)

56. The line $y + 4 = 0$ is equivalent to $y = -4$, which is the equation of a horizontal line. Any line perpendicular to it must be a vertical line. Thus the required equation is $x = -3$. (**Writing Equations of Lines**)

57.
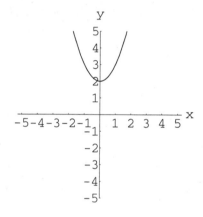
(**Graphing**)

58.
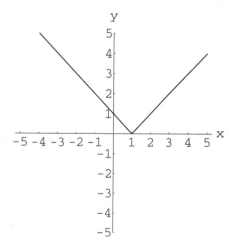
(**Graphing**)

59.   **(Graphing)**

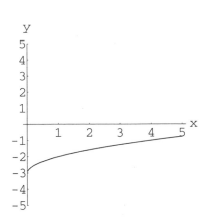

60.   **(Graphing)**

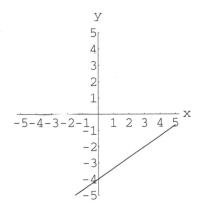

61. First write the equation in slope-intercept form:

$$-3y = -2x + 6 \quad \text{Isolate the } y\text{-term}$$

$$y = \frac{2}{3}x - 2 \quad \text{Divide each term by } -3$$

Now, the y-intercept is -2 and the slope is $\dfrac{\text{rise}}{\text{run}} = \dfrac{2}{3}$. Begin with a point at $(0, -2)$ and then rise 2 units, and run to the right 3 units to the point $(3, 0)$. Draw the line that contains both of these points:

                                  (Graphing)

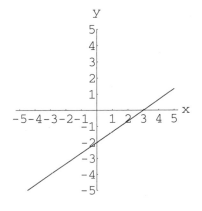

62. 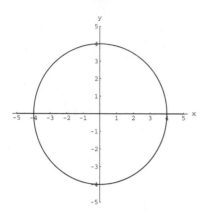 **(Graphing)**

63. 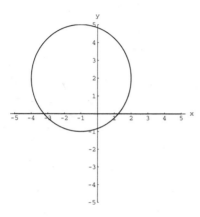 **(Graphing)**

64. Note that this is a parabola that has been shifted up 3 units and to the right 1 unit. The parabola opens up because $a = 1$.

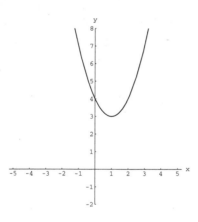 **(Graphing)**

65. 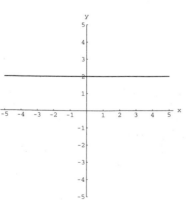 **(Graphing)**

66. So that $x^2 + y^2 + 4x - 2y = 10$ can be graphed easily, the equation must be put in standard form for a circle. To put this in standard form, complete the square on the x terms and on the y terms:

$$x^2 + 4x + 4 + y^2 - 2y + 1 = 10 + 4 + 1$$

$$(x+2)^2 + (y-1)^2 = 15$$

This circle has a center at $(-2, 1)$ and a radius of $\sqrt{15} \approx 3.9$. On the graph, use a radius slightly less than 4:

(Graphing)

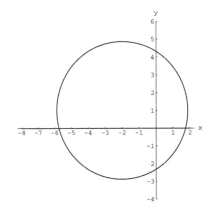

67. To test for *symmetry with respect to the x-axis*, replace y with $(-y)$:

$$x^2 + (-y)^2 = 1$$

$$x^2 + y^2 = 1$$

This equation is equivalent to the original equation which implies that the graph is symmetric with respect to the x-axis.

Symmetry with respect to the y-axis:

$$(-x)^2 + y^2 = 1$$

$$x^2 + y^2 = 1$$

Since replacing x with $(-x)$ gave an equation equivalent to the original equation, the graph is symmetric with respect to the y-axis.

Symmetry with respect to the origin:

$$(-x)^2 + (-y)^2 = 1$$

$$x^2 + y^2 = 1$$

Since replacing both x with $(-x)$ and y with $(-y)$ gave an equation equivalent to the original equation, the graph is symmetric with respect to the origin. Note that $x^2 + y^2 = 1$ is the equation of a circle with center $(0, 0)$ and hence must be symmetric to each axis and the origin. **(Symmetry)**

68. *Symmetry with respect to the x-axis*:

$$-y = \frac{1}{2}x + 4$$

$y = -\frac{1}{2}x - 4$ is not equivalent to the original equation. Therefore the graph is *not* symmetric with respect to the *x*-axis.

Symmetry with respect to the y-axis:

$$y = \frac{1}{2}(-x) + 4$$

$y = -\frac{1}{2}x + 4$. Not equivalent, not symmetric with respect to the *y*-axis.

Symmetry with respect to the origin:

$$(-y) = \frac{1}{2}(-x) + 4$$

$y = \frac{1}{2}x - 4$. Not equivalent, not symmetric with respect to the origin. **(Symmetry)**

69. *Symmetry with respect to the x-axis:*

$$-y = x^2 - 3$$

$y = -x^2 + 3$. Not equivalent, not symmetric with respect to the *x*-axis.

Symmetry with respect to the y-axis:

$$y = (-x)^2 - 3$$

$y = x^2 - 3$. This is equivalent, so graph is symmetric with respect to the *y*-axis.

Symmetry with respect to the origin:

$$-y = (-x)^2 - 3$$

$$-y = x^2 - 3$$

$y = -x^2 + 3$. Not equivalent, not symmetric with respect to the origin. **(Symmetry)**

70. *Symmetry with respect to the x-axis:*

$$-y = x^5 + x^3$$

$y = -x^5 - x^3$. Not equivalent, not symmetric with respect to the *x*-axis.

Symmetry with respect to the y-axis:

$$y = (-x)^5 + (-x)^3$$

$y = -x^5 - x^3$. Not equivalent, not symmetric with respect to the y-axis.

Symmetry with respect to the origin:

$$-y = (-x)^5 + (-x)^3$$

$y = x^5 + x^3$. This is equivalent, so the graph is symmetric with respect to the origin. **(Symmetry)**

71. Notice that only one of the variables is raised to the 2nd power. This is in the form of a parabola. **(Graphing)**

72. Notice that both the x and y variables are raised to the 1st power. This is in the standard form of a line. **(Graphing)**

73. Because both the x and y are raised to the 2nd power, and the coefficients of the squared terms are positive and equal, and the constant on the right side is positive, the graph of this equation is a circle. Although this is not in the standard form of a circle, we could complete the square on the x and y terms to put it in standard form. **(Graphing)**

74. The only variable is raised to the 1st power. This is the equation of a vertical line. **(Graphing)**

75. The only variable is raised to the 1st power. This is the equation of a horizontal line. **(Graphing)**

Grade Yourself

Circle the question numbers that you had incorrect. Then indicate the number of questions you missed. If you answered more than three questions incorrectly, you will then have to focus on that topic. If a topic has less than three questions and you had at least one wrong, we suggest you study that topic also. Read your textbook, a review book, or ask your teacher for help.

Subject: Algebra Review

Topic	Question Numbers	Number Incorrect
Real Numbers	1, 2	
Interval Notation	3, 4, 5, 6, 7, 8, 9	
Solving First-degree and Compound Inequalities	10, 11, 12, 13, 14, 15	
Absolute Value	16, 17, 18, 19, 20, 21, 22, 23, 24	
Quadratic Inequalities	25, 26, 27	
The Distance and Midpoint Formulas	28, 29, 30, 31, 32, 33	
Finding the Equation of a Circle	34, 35, 36, 37	
Finding the Center and Radius of a Circle	38, 39, 40, 41	
Slope	42, 43, 44, 45	
Writing Equations of Lines	46, 47, 48, 49, 50, 51, 52, 53, 54, 55, 56	
Graphing	57, 58, 59, 60, 61, 62, 63, 64, 65, 66, 71, 72, 73, 74, 75	
Symmetry	67, 68, 69, 70	

Functions and Limits

Brief Yourself

This chapter contains information about functions, including how to determine when a graph or set of ordered pairs defines a function, how to find the domain and range of a function, and how to sketch graphs of functions. The algebra of functions and trigonometric functions is also reviewed. The limit questions include how to find the limit given a graph, or an algebraic or trigonometric expression. The chapter concludes with questions about the continuity of functions.

A function is a set of ordered pairs where each x-coordinate is paired with a unique y-coordinate. This leads to the vertical line test for the graph of a function—that is, if a vertical line passes through more than one point on the graph of an equation, that equation does *not* represent the equation of a function.

When looking for the domain of a function, remember that 0 is not allowed as a denominator, and negative numbers under square roots (or any other even root) are also not defined. By setting the denominator of a fraction equal to 0 and solving, we can determine which values of the variable to eliminate from the domain. For equations involving square roots (or any even roots), we set the radicand greater than or equal to 0, solve, and use the result for the domain.

The algebra of functions includes the general operations of addition, subtraction, multiplication, and division of functions. When we perform $(f \circ g)(x)$, the composition of f and g, we first find $g(x)$ and then use this result in f; that is, $(f \circ g)(x) = f(g(x))$.

A thorough review of trigonometry is necessary to succeed in calculus. If your textbook doesn't provide a sufficient review, check the bookstore or library for a trigonometry book. Make and use flashcards to help you memorize the Pythagorean Identities and other Fundamental Identities. A brief list of these identities follows:

$$\sin^2\theta + \cos^2\theta = 1 \qquad 1 + \tan^2\theta = \sec^2\theta \qquad 1 + \cot^2\theta = \csc^2\theta$$

$$\sin\theta = \frac{1}{\csc\theta} \qquad \cos\theta = \frac{1}{\sec\theta} \qquad \tan\theta = \frac{1}{\cot\theta} = \frac{\sin\theta}{\cos\theta}$$

$$\sin(-\theta) = -\sin\theta \qquad \cos(-\theta) = \cos\theta \qquad \tan(-\theta) = -\tan\theta$$

$$\sin 2\theta = 2\sin\theta\cos\theta \qquad \cos 2\theta = \cos^2\theta - \sin^2\theta$$

To find the limit of $f(x)$ as x approaches c, try substituting c into the given expression. If this produces an undefined expression (such as 0 in the denominator), try some algebraic techniques such as factoring, rationalizing the numerator or denominator, combining fractions, etc., to help you simplify the given expression. If you are given a graph and asked to find a limit as x approaches c, remember that a limit can exist even if the graph does not contain a y value at c.

For questions concerning continuity, remember that polynomial functions, rational functions, radical functions, and trigonometric functions are continuous at every point in their *domain*. Therefore, a rational function that contains a vertical asymptote at $x = c$ cannot be continuous at c because c would not be in the domain of that rational function.

Test Yourself

Identify which of the following are the graphs of functions.

1.

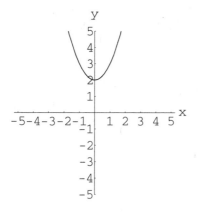

2.

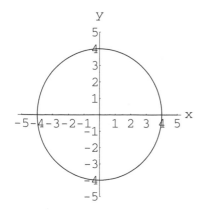

3.

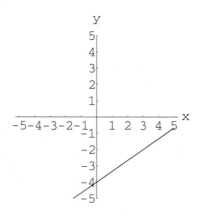

4. Determine whether the given set defines a function: $\{(0, 1), (1, 2), (2, 3)\}$

5. Determine whether the given set defines a function: $\{1, 2, 3\}$

6. Determine whether the given set defines a function: $\{(2, 3), (2, -3), (3, 4)\}$

7. Find the domain: $y = \frac{1}{2}x - 3$

8. Find the domain: $y = \frac{x + 1}{x - 4}$

9. Find the domain: $y = \sqrt{2x + 1}$

10. Find the domain: $y = \sqrt{x^2 - x - 6}$

11. Find the domain: $y = \dfrac{\sqrt{3-x}}{x-1}$

12. Given $f(x) = -x^2 - 2x$, find $f(-1)$

13. Given $g(x) = 2x^2 + 3x + 1$, find $g(x+h)$

14. Given $h(x) = \dfrac{2x+5}{x-3}$, find $h(0)$

15. Sketch the graph of $f(x) = x^2 - 4$

16. Sketch the graph of $f(x) = (x+2)^3 - 1$

17. Sketch the graph of $f(x) = -|x-3| + 4$

18. Use the graph of $f(x) = \sqrt{x}$ to determine a formula for the given function:

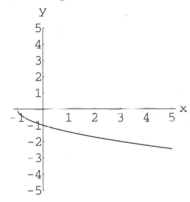

19. If $f(x) = 3x^2 + 1$ and $g(x) = 4x - 3$, find:

    a) $(f+g)(x)$

    b) $(f-g)(2)$

    c) $(fg)(x)$

    d) $\left(\dfrac{f}{g}\right)(-2)$

20. If $f(x) = 3x + 4$ and $g(x) = 2x - 5$, find $(fg)(x)$ and $(f/g)(x)$

21. If $f(x) = 6x - 1$ and $g(x) = 5 + 3x$, find $(f \circ g)(x)$

22. If $f(x) = 3x^2 - 2x$ and $g(x) = 2x - 5$, find $(f \circ g)(-1)$.

23. If $g(x) = 2x + 7$ and $h(x) = x^2$, find $(g \circ h)(x)$ and $(h \circ g)(x)$. Is compostion of functions commutative?

24. Given $\sin\theta = \dfrac{2}{5}$, find $\csc\theta$

25. Given: Find the six trigono-metric functions of θ, θ in Quadrant I

26. If $\cot\theta = \dfrac{5}{4}$, find $\csc\theta$

27. Complete the table with *exact* values.

	$\dfrac{\pi}{3}$	$\dfrac{\pi}{2}$	$\dfrac{5\pi}{6}$
sin			
cos			
tan			

28. Express $225°$ in radian measure as a multiple of π

29. Express $\dfrac{7\pi}{6}$ in degrees

30. Solve $4\cos^2\theta - 1 = 0$ for θ, $0 \le \theta < 2\pi$

31. Solve $2\sin\theta + \sqrt{3} = 0$ for θ, $0 \le \theta < 2\pi$

32. Solve $\tan^2\theta - 1 = 0$ for θ, $0 \le \theta < 2\pi$

33. Prove $\lim\limits_{x \to -2} (3x+1) = -5$

Evaluate each limit.

34. $\lim\limits_{x \to 2} (x^2 - 4)$

35. $\lim\limits_{x \to -3} \left(\dfrac{x-3}{2x+1}\right)$

36. $\lim\limits_{x \to 3} \left(\dfrac{x^2 - 9}{x-3}\right)$

37. $\lim_{x \to 0} \left(\dfrac{\sqrt{x + 16} - 4}{x} \right)$

38. $\lim_{x \to \frac{\pi}{2}} \cos x$

39. $\lim_{x \to 0} \dfrac{\sin 4x}{x}$

40. $\lim_{x \to 0} \dfrac{\sin 2x}{\sin 3x}$

41. Use the given graph of f to find each limit.

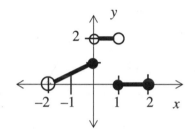

a) $\lim_{x \to -2^+} f(x)$

b) $\lim_{x \to 0^-} f(x)$

c) $\lim_{x \to 0^+} f(x)$

d) $\lim_{x \to 0} f(x)$

e) $\lim_{x \to 1^+} f(x)$

42. $\lim_{x \to 2^-} (4 - 3x)$

43. $\lim_{x \to 3^+} \dfrac{|x - 3|}{x - 3}$

44. $\lim_{x \to 3^-} \dfrac{|x - 3|}{x - 3}$

45. $\lim_{x \to 2^-} \sqrt{x - 2}$

46. Determine whether $f(x) = \dfrac{2}{3}x - 1$ is continuous at $x = 4$

47. Determine whether $f(x) = \dfrac{x^2 - 4}{x - 2}$ is continuous at $x = 2$

48. Determine whether $f(x) = \begin{cases} x^2 - 2, x \neq 0 \\ 0, x = 0 \end{cases}$ is continuous at $x = 0$

49. Determine whether $f(x) = \dfrac{x - 1}{x^2 + x - 12}$ is continuous at $x = 2$

 # Check Yourself

1. This is a function. Note that any vertical line passes through at most one point on the graph. (**Function**)

2. Not a function; doesn't pass the vertical line test. (**Function**)

3. Function. (**Function**)

4. Function—no x-coordinate is repeated. (**Function**)

5. Not a function. This is not a set of ordered pairs. (**Function**)

6. Not a function. The x-coordinate 2 is paired with more than one y-coordinate. (**Function**)

7. $(-\infty, \infty)$, or all real numbers. (**Domain and Range**)

8. $(-\infty, 4) \cup (4, \infty)$, or $\{x \mid x \neq 4\}$. When $x = 4$, the denominator equals 0, which is not allowed. **(Domain and Range)**

9. To be defined, the radicand, $2x + 1$, must be greater than or equal to 0:

    $2x + 1 \geq 0$ Set the radicand ≥ 0

    $2x \geq -1$ Subtract 1 from both sides of the inequality

    $x \geq -\dfrac{1}{2}$ Divide both sides of the inequality by 2

    Written in interval notation, the domain is $[-\frac{1}{2}, \infty)$. Written in set notation, the domain is $\{x \mid x \geq -\frac{1}{2}\}$, **(Domain and Range)**

10. To be defined, the radicand must be greater than or equal to 0:

    $x^2 - x - 6 \geq 0$ Set radicand ≥ 0

    $(x - 3)(x + 2) \geq 0$ Factor

    Find the split points by setting each factor equal to 0 and solving for x. Therefore, 3 and –2 are the split points.

    ```
    x - 3      −        −        +

    x + 2      −        +        +
           ┼─────────┼─────────┼
                   -2        3
    (x - 3) (x + 2)   ⊕        ⊖        ⊕
    ```

    Since the radicand must be greater than or equal to 0, the domain is $(-\infty, -2] \cup [3, \infty)$, or $\{x \mid x \leq -2 \text{ or } x \geq 3\}$. **(Domain and Range)**

11. Since the denominator cannot equal 0, $x \neq 1$. Since the radicand must be greater than or equal to 0

    $3 - x \geq 0$ Set the radicand ≥ 0

    $x \leq 3$ Solve for x

    Putting both of these requirements together, the domain is $(-\infty, 1) \cup (1, 3]$, or $\{x \mid x \leq 3, x \neq 1\}$. **(Domain and Range)**

12. $f(-1)$ means substitute -1 for x in f:

    $f(-1) = -(-1)^2 - 2(-1) = -1 + 2 = 1$ **(Function Notation)**

13. $g(x + h)$ means substitute $(x + h)$ for x in g:

    $g(x + h) = 2(x + h)^2 + 3(x + h) + 1$ Replace x with $(x + h)$

    $= 2(x^2 + 2xh + h^2) + 3x + 3h + 1$ Expand $(x + h)^2$

    $= 2x^2 + 4xh + 2h^2 + 3x + 3h + 1$ **(Function Notation)**

14. $h(0) = \dfrac{2(0) + 5}{0 - 3} = -\dfrac{5}{3}$ **(Function Notation)**

15. The basic graph has the shape of a parabola opening up. The –4 means the graph is shifted down 4 units:

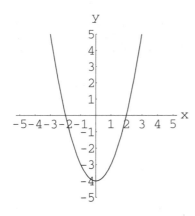

(Graphs of Functions)

16. The basic graph is a cubic. The + 2 in the parentheses with x means the graph is shifted 2 units to the left. The –1 means the graph is shifted down 1 unit:

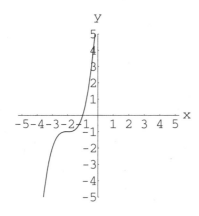

(Graphs of Functions)

17. An absolute value graph is a v-shaped graph. The –3 means the graph is shifted 3 units to the right. The + 4 means the graph is shifted up 4 units. The negative in front of the absolute value bars means the v opens down:

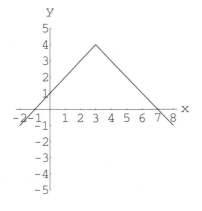

(Graphs of Functions)

18. The graph is shifted 1 unit to the left, and reflected across the x-axis, so $f(x) = -\sqrt{x + 1}$. **(Graphs of Functions)**

19. a) $(f + g)(x) = (3x^2 + 1) + (4x - 3) = 3x^2 + 4x - 2$

b) $(f - g)(x) = (3x^2 + 1) - (4x - 3) = 3x^2 - 4x + 4$

Now, $(f - g)(2) = 3(2)^2 - 4(2) + 4 = 8$

Note that the same result is obtained by finding $f(2) - g(2)$

c) $(fg)(x) = (3x^2 + 1)(4x - 3) = 12x^3 - 9x^2 + 4x - 3$

d) $\left(\dfrac{f}{g}\right)(-2) = \dfrac{3(-2)^2 + 1}{4(-2) - 3} = \dfrac{13}{-11} = -\dfrac{13}{11}$ **(Operations with Functions)**

20. $(fg)(x) = (3x + 4)(2x - 5) = 6x^2 - 15x + 8x - 20 = 6x^2 - 7x - 20$

$(f/g)(x) = \dfrac{3x + 4}{2x - 5}$. You cannot use cancellation here to simplify because the numerator and denominator

of the fraction contain *terms*, not *factors*. **(Operations with Functions)**

21. $(f \circ g)(x) = f(g(x))$ Use the definition of composition of functions

$= f(5+3x)$ Since $g(x) = 5 + 3x$, substitute $5 + 3x$ for $g(x)$

$= 6(5+3x) - 1$ $f(5 + 3x)$ means substitute $5 + 3x$ for x in f

$= 30 + 18x - 1$ Distribute

$= 18x + 29$ Combine like terms **(Composition of Functions)**

22. $(f \circ g)(-1) = f(g(-1))$ Use the definition of composition of functions

$= f(2(-1) - 5)$ Find $g(-1)$

$= f(-7)$ Now substitute -7 for x in f

$= 3(-7)^2 - 2(-7) = 161$ **(Composition of Functions)**

23. $(g \circ h)(x) = g(h(x)) = g(x^2) = 2(x^2) + 7 = 2x^2 + 7$

$(h \circ g)(x) = h(g(x)) = h(2x + 7) = (2x + 7)^2 = 4x^2 + 28x + 49$

No, composition of functions is not commutative. For example, in this problem, $(g \circ h)(x) \neq (h \circ g)(x)$.

(Composition of Functions)

24. Since $\csc\theta = \dfrac{1}{\sin\theta}$, $\csc\theta = \dfrac{1}{2/5} = \dfrac{5}{2}$ **(Trigonometric Function Definitions)**

25. Use the Pythagorean theorem to find the missing side:

$4^2 = b^2 + 3^2$, $7 = b^2$, $\sqrt{7} = b$

$\sin\theta = \dfrac{\text{side opposite}}{\text{hypotenuse}} = \dfrac{\sqrt{7}}{4}$

$$\cos \theta = \frac{\text{side adjacent}}{\text{hypotenuse}} = \frac{3}{4}$$

$$\tan \theta = \frac{\text{opposite}}{\text{adjacent}} = \frac{\sqrt{7}}{3}$$

$$\csc \theta = \frac{4}{\sqrt{7}}$$

$$\sec \theta = \frac{4}{3}$$

$$\cot \theta = \frac{3}{\sqrt{7}} \quad \textbf{(Trigonometric Function Definitions)}$$

26. Use the Pythagorean Identity:

$$1 + \cot^2 \theta = \csc^2 \theta$$

$$1 + \left(\frac{5}{4}\right)^2 = \csc^2 \theta \qquad \text{Use } \cot \theta = \frac{5}{4}$$

$$1 + \frac{25}{16} = \csc^2 \theta \qquad \text{Square } \frac{5}{4}$$

$$\frac{41}{16} = \csc^2 \theta \qquad 1 + \frac{25}{16} = \frac{16}{16} + \frac{25}{16} = \frac{41}{16}$$

$$\pm \frac{\sqrt{41}}{4} = \csc \theta \qquad \text{Take the square root of both sides of the equation} \quad \textbf{(Trigonometric Function Definitions)}$$

27. The values requested for this table should be memorized:

Angle	$\frac{\pi}{3}$	$\frac{\pi}{2}$	$\frac{5\pi}{6}$
sin	$\frac{\sqrt{3}}{2}$	1	$\frac{1}{2}$
cos	$\frac{1}{2}$	0	$-\frac{\sqrt{3}}{2}$
tan	$\sqrt{3}$	undefined	$-\frac{\sqrt{3}}{3}$

(Special Angles)

28. Set up a proportion:

$$\frac{225°}{x} = \frac{180°}{\pi}$$

$180x = 225\pi$ Cross-multiply

$x = \dfrac{225}{180}\pi$ Divide both sides by 180

$x = \dfrac{5\pi}{4}$ Reduce the fraction. Note that this can also be written as $\dfrac{5}{4}\pi$. **(Degrees and Radians)**

29. Although we could set up a proportion as we did in #28, it is usually easier to use $\pi = 180°$ to write:

$\dfrac{7\pi}{6} = \dfrac{7(180°)}{6} = 210°$ **(Degrees and Radians)**

30. $4\cos^2\theta - 1 = 0$ Given

$(2\cos\theta - 1)(2\cos\theta + 1) = 0$ Factor the left side as a difference of two squares

$2\cos\theta - 1 = 0$ or $2\cos\theta + 1 = 0$ Solve each equation

$\cos\theta = \dfrac{1}{2}$ 　　　　$\cos\theta = -\dfrac{1}{2}$ These values should be memorized (they are special angles). However, you can also use the $\cos^{-1}$ on your calculator to find the reference angle of $\pi/3$. We need a reference angle of $\pi/3$ in all four quadrants (because we want $\cos\theta = \pm\dfrac{1}{2}$), so use $\pi/3$, $\pi - \pi/3$, $\pi + \pi/3$ and $2\pi - \pi/3$ to find:

$\theta = \dfrac{\pi}{3}, \dfrac{5\pi}{3}$ 　　　　$\theta = \dfrac{2\pi}{3}, \dfrac{4\pi}{3}$

(Solving Trigonometric Equations)

31. $2\sin\theta + \sqrt{3} = 0$ Given

$2\sin\theta = -\sqrt{3}$ Subtract $-\sqrt{3}$ from both sides of the equation

$\sin\theta = -\dfrac{\sqrt{3}}{2}$ Divide both sides of the equation by 2

The reference angle is $\dfrac{\pi}{3}$. The sine function is negative in quadrants III and IV. Therefore, $\theta = \dfrac{4\pi}{3}, \dfrac{5\pi}{3}$.
(Solving Trigonometric Equations)

32. $\tan^2\theta - 1 = 0$ Given

$\tan^2\theta = 1$ Add 1 to both sides of the equation

$\tan\theta = \pm 1$ Take the square root of both sides of the equation

The reference angle is $\dfrac{\pi}{4}$, and there is a solution in each of the four quadrants. $\theta = \dfrac{\pi}{4}, \dfrac{3\pi}{4}, \dfrac{5\pi}{4}, \dfrac{7\pi}{4}$ **(Solving Trigonometric Equations)**

33. Prove that for each $\varepsilon > 0$, there exists a $\delta > 0$ such that if $0 < |x - (-2)| < \delta$ then

    $|(3x + 1) - (-5)| < \varepsilon$.

    Simplify: $|(3x + 1) - (-5)| = |3x + 6| = |3| \, |x + 2| = 3|x + 2|$. Now,

    $|(3x + 1) - (-5)| < \varepsilon$ is equivalent to $3|x + 2| < \varepsilon$, which is equivalent to $|x + 2| < \dfrac{\varepsilon}{3}$.

    Proof:

    Let $\delta = \dfrac{\varepsilon}{3}$

    $|x + 2| < \delta \Rightarrow |x + 2| < \dfrac{\varepsilon}{3} \Rightarrow 3|x + 2| < \varepsilon \Rightarrow |3x + 6| < \varepsilon \Rightarrow |(3x + 1) - (-5)| < \varepsilon$

    ($\varepsilon - \delta$ Definition of a Limit)

34. We can use substitution: $\lim\limits_{x \to 2} (x^2 - 4) = 2^2 - 4 = 0$ **(Evaluating Limits)**

35. $\lim\limits_{x \to -3} \dfrac{x - 3}{2x + 1} = \dfrac{-3 - 3}{2(-3) + 1} = \dfrac{-6}{-5} = \dfrac{6}{5}$ **(Evaluating Limits)**

36. Substitution will not work because it causes division by 0. Factor:

    $\lim\limits_{x \to 3} \dfrac{x^2 - 9}{x - 3} = \lim\limits_{x \to 3} \dfrac{(x - 3)(x + 3)}{x - 3} = \lim\limits_{x \to 3} (x + 3) = 6$

    (Evaluating Limits)

37. Substitution will not work because it causes division by 0. Rationalize the numerator by multiplying by the conjugate of the numerator:

    $\lim\limits_{x \to 0} \dfrac{\sqrt{x + 16} - 4}{x} \cdot \dfrac{\sqrt{x + 16} + 4}{\sqrt{x + 16} + 4}$ Multiply by the conjugate/conjugate

    $= \lim\limits_{x \to 0} \dfrac{x + 16 - 16}{x(\sqrt{x + 16} + 4)}$ Multiply the numerators

    $= \lim\limits_{x \to 0} \dfrac{x}{x(\sqrt{x + 16} + 4)}$ Combine similar terms in the numerator

    $= \lim\limits_{x \to 0} \dfrac{1}{\sqrt{x + 16} + 4}$ Use $\dfrac{x}{x} = 1$

    $= \dfrac{1}{\sqrt{16} + 4} = \dfrac{1}{8}$ Let $x = 0$ and simplify the denominator

    (Evaluating Limits)

38. Use substitution: $\lim\limits_{x \to \frac{\pi}{2}} \cos x = \cos(\pi/2) = 0$ **(Trigonometric Limits)**

39. Use $\lim\limits_{x \to 0} \dfrac{\sin x}{x} = 1$

$$\lim\limits_{x \to 0} \frac{\sin 4x}{x} \cdot \frac{4}{4}$$

$$= 4 \lim\limits_{x \to 0} \frac{\sin 4x}{4x} = 4(1) = 4$$

(Trigonometric Limits)

40. $\lim\limits_{x \to 0} \dfrac{\sin 2x}{\sin 3x} = \lim\limits_{x \to 0} \dfrac{\sin 2x}{1} \cdot \dfrac{1}{\sin 3x}$

$$= \lim\limits_{x \to 0} \frac{2x}{2x} \cdot \frac{\sin 2x}{1} \cdot \frac{1}{\sin 3x} \cdot \frac{3x}{3x}$$

$$= \lim\limits_{x \to 0} \frac{2x}{3x} \cdot \frac{\sin 2x}{2x} \cdot \frac{3x}{\sin 3x}$$

$$= \lim\limits_{x \to 0} \frac{2}{3} \cdot 1 \cdot 1 = \frac{2}{3}$$

(Trigonometric Limits)

41. a) 0

    b) 1

    c) 2

    d) Does not exist (since b and c do not have the same answer)

    e) 0 **(One-sided Limits)**

42. Use substitution: $\lim\limits_{x \to 2^-} (4 - 3x) = 4 - 3(2) = -2$ **(One-sided Limits)**

43. As x approaches 3 from the right, $x - 3 > 0 \Rightarrow |x - 3| = x - 3$. Therefore,

$$\lim\limits_{x \to 3^+} \frac{|x - 3|}{x - 3} = \lim\limits_{x \to 3^+} \frac{x - 3}{x - 3} = 1 \quad \textbf{(One-sided Limits)}$$

44. As x approaches 3 from the left, $x - 3 < 0 \Rightarrow |x - 3| = -(x - 3)$. Therefore,

$$\lim\limits_{x \to 3^-} \frac{|x - 3|}{x - 3} = -1 \quad \textbf{(One-sided Limits)}$$

45. Substitution would lead to an incorrect answer of 0. The graph of $f(x) = \sqrt{x-2}$ does *not* exist to the left of 2. Therefore, $\lim\limits_{x \to 2^-} \sqrt{x-2}$ does not exist. **(One-sided Limits)**

46. Continuous. Linear functions are continuous everywhere. **(Continuity)**

47. Not continuous at $x = 2$ since when $x = 2$ the denominator equals 0. **(Continuity)**

48. Not continuous. $\lim\limits_{x \to 0} f(0) = -2$, but $f(0) = 0$. **(Continuity)**

49. Continuous. A rational function is continuous on its domain. The domain of f is all real numbers except 3 and -4. **(Continuity)**

Grade Yourself

Circle the question numbers that you had incorrect. Then indicate the number of questions you missed. If you answered more than three questions incorrectly, you will then have to focus on that topic. If a topic has less than three questions and you had at least one wrong, we suggest you study that topic also. Read your textbook, a review book, or ask your teacher for help.

Subject: *Functions and Limits*

Topic	Question Numbers	Number Incorrect
Functions	1, 2, 3, 4, 5, 6	
Domain and Range	7, 8, 9, 10, 11	
Function Notation	12, 13, 14	
Graphs of Functions	15, 16, 17, 18	
Operations with Functions	19, 20	
Composition of Functions	21, 22, 23	
Trigonometric Function Definitions	24, 25, 26	
Special Angles	27	
Degrees and Radians	28, 29	
Solving Trigonometric Equations	30, 31, 32	
ε-δ Definition of a Limit	33	
Evaluating Limits	34, 35, 36, 37	
Trigonometric Limits	38, 39, 40	
One-sided Limits	41, 42, 43, 44, 45	
Continuity	46, 47, 48, 49	

Derivatives

3

Brief Yourself

This chapter contains test items for derivatives. There are items to test your knowledge of the limit definition of the derivative, as well as items where you will use the power rule, the sum and difference rules, and the product and quotient rules to find derivatives. This chapter also contains items to test your knowledge of the chain rule for algebraic and trigonometric functions.

If you are asked to use the limit definition of the derivative to find a derivative,

$$(f'(x) = \lim_{\Delta x \to 0} \frac{f(x + \Delta x) - f(x)}{\Delta x}$$, for example) try to check your answer by using the other rules learned in the chapter (sum and difference rules, product and quotient rules).

Be aware of the different notations for derivatives. This will help you recognize the problems requiring higher-order derivatives. Some of the notations you may see for the first derivative include: y', $f'(x)$, $\frac{dy}{dx}$, $D_x[y]$, $\frac{d}{dx}[f(x)]$. Each of these notations can be changed to indicate higher-order derivatives. For example, each of the following indicates the third derivative: y''', $f'''(x)$, $\frac{d^3 y}{dx^3}$, $D_x^3[y]$, $\frac{d^3}{dx^3}[f(x)]$. If you are asked to find a higher-order derivative such as $y^{(50)}$, examine the first few derivatives, and look for a pattern.

Make and use flashcards to help you memorize the derivatives of the trigonometric functions. A list of the first derivatives of the trigonometric functions follows:

$$\frac{d}{dx}[\sin u] = (\cos u)\, u'$$

$$\frac{d}{dx}[\cos u] = (-\sin u)\, u'$$

$$\frac{d}{dx}[\tan u] = \left(\sec^2 u\right) u'$$

$$\frac{d}{dx}[\csc u] = (-\csc u \cot u)\, u'$$

$$\frac{d}{dx}[\sec u] = (\sec u \tan u)\, u'$$

$$\frac{d}{dx}[\cot u] = \left(-\csc^2 u\right) u'$$

Implicit differentiation problems will also be presented. Although the directions may include "use implicit differentiation," the student should be able to recognize the types of problems that require this technique—those that involve equations that are not solved for y or $f(x)$.

Remember that many different types of problems involve finding derivatives. To find the equation of a tangent line to the graph of a function $f(x)$, you must find the slope of that tangent line, and that is the derivative, $f'(x)$. If you are asked about the velocity (implying instantaneous velocity), you are also being asked to find a derivative.

Test Yourself

Use the limit definition,

$$f'(x) = \lim_{\Delta x \to 0} \frac{f(x + \Delta x) - f(x)}{\Delta x}, \text{ to find each}$$

derivative.

1. $f(x) = 5 - 2x$

2. $f(x) = x^2 + 4x$

3. $f(x) = \dfrac{4}{x - 1}$

Use the limit definition, $f'(x) =$

$$\lim_{\Delta x \to 0} \frac{f(x + \Delta x) - f(x)}{\Delta x}, \text{ to find the slope of}$$

the tangent to the graph of $f(x)$ at the given point. Check your answer using the differentiation rules.

4. $f(x) = 2x - 3$ at $(-1, -5)$

5. $f(x) = x^2 - 2x + 1$ at $(2, 1)$

6. Use the limit definition of velocity to find the velocity after 3 seconds of an object traveling along a line whose distance traveled after t seconds is $s(t) = \sqrt{3t + 8}$.

Find $f'(x)$ for each function.

7. $f(x) = \pi$

8. $f(x) = 6x^2 - 3x + 1$

9. $f(x) = \dfrac{2}{\sqrt[3]{x}} + \sqrt{x}$

10. Find the equation of the tangent line to the graph of $f(x) = 3x^2 - 2x + 4$ at $x = 2$.

Find $\dfrac{dy}{dx}$ for each function.

11. $y = (3x + 2)(x^2 - 4)$

12. $y = \dfrac{4x + 5}{2x - 1}$

Find the derivative.

13. $y = (3x^2 + 2x)^3$

14. $f(x) = 4x(x^2 - 6x)^2$

15. $h(x) = \left(\dfrac{2x - 3}{x + 1}\right)^4$

16. $g(x) = \sqrt{x^2 + 4x - 1}$

Find y'.

17. $y = \sin(4x + 6)$

18. $y = \cos(x^2)$

19. $y = 3\tan 2x$

20. $y = \csc(2x^2 - 3)$

21. $y = -2\sec \pi x$

22. Find $D_x\left(4\cot\left(x + \dfrac{\pi}{2}\right)\right)$

Differentiate.

23. $y = 4x^2\cos 2x$

24. $y = \sin x \cot x$

25. $y = \dfrac{\sin x}{1 + \tan x}$

26. $y = \sin^2 2x$

27. $y = \tan^3(3x + 1)$

28. $y = \csc 2x \sec 3x$

29. $y = \cos^4(3x + 1)$

30. Find $D_x\sqrt{\sin x}$

31. Find $\dfrac{d^2y}{dx^2}$ if $y = x\sin x$

32. Find $\dfrac{d^2y}{dx^2}$ if $y = \cos^2 x$

33. Find $\dfrac{d^{50}y}{dx^{50}}$ if $y = \sin x$

34. Find $D_x^{\,2}(\tan 2x)$

35. Find y''' if $y = 6x^3 + 8x^2 - 5x + 1$

36. Find y'' if $y = (2x + 5)^4$

37. Find y' if $x^2 + 2xy + y^2 = 6$

38. Find y' if $x^2y^2 - 3x + 10 = 0$

39. Find y' if $x^2 + \sin y = 4$

40. Find y' if $\cos xy = 4x^2 + 1$

41. Find y' if $y\tan x - x^2 = 6$

42. Find the equation of the tangent line to
$4x^2 + y^2 = 100$ at $(3, 8)$

43. Find the equation of the tangent line to
$x^2 + 4y^2 - 2x = 3$ at $\left(2, \dfrac{\sqrt{3}}{2}\right)$

Check Yourself

1. $\displaystyle\lim_{\Delta x \to 0} \dfrac{5 - 2(x + \Delta x) - (5 - 2x)}{\Delta x}$ Replace x with $(x + \Delta x)$ in f to get the $f(x + \Delta x)$ term.

$= \displaystyle\lim_{\Delta x \to 0} \dfrac{5 - 2x - 2\Delta x - 5 + 2x}{\Delta x}$ Use the distributive property to remove parentheses

$= \displaystyle\lim_{\Delta x \to 0} \dfrac{-2\Delta x}{\Delta x}$ Combine similar terms

$= -2$ Since $\Delta x \neq 0$, we can reduce using $\dfrac{\Delta x}{\Delta x} = 1$

(Limit Definition of Derivative)

2. $\lim\limits_{\Delta x \to 0} \dfrac{(x + \Delta x)^2 + 4(x + \Delta x) - (x^2 + 4x)}{\Delta x}$ Replace x with $(x + \Delta x)$ in f

$= \lim\limits_{\Delta x \to 0} \dfrac{x^2 + 2x\Delta x + (\Delta x)^2 + 4x + 4\Delta x - x^2 - 4x}{\Delta x}$ Expand $(x + \Delta x)^2$ and use the distributive property to remove parentheses

$= \lim\limits_{\Delta x \to 0} \dfrac{2x\Delta x + (\Delta x)^2 + 4\Delta x}{\Delta x}$ Combine similar terms

$= \lim\limits_{\Delta x \to 0} \dfrac{\Delta x\,(2x + \Delta x + 4)}{\Delta x}$ Factor out Δx from the numerator

$= \lim\limits_{\Delta x \to 0} (2x + \Delta x + 4)$ Since $\Delta x \neq 0$, we can reduce using $\dfrac{\Delta x}{\Delta x} = 1$

$= 2x + 4$ Substitute $\Delta x = 0$ to complete the limit **(Limit Definition of Derivative)**

3. $\lim\limits_{\Delta x \to 0} \dfrac{\dfrac{4}{x + \Delta x - 1} - \dfrac{4}{x - 1}}{\Delta x}$ Replace x with $(x + \Delta x)$ in f

$= \lim\limits_{\Delta x \to 0} \dfrac{\dfrac{4x - 4 - 4x - 4\Delta x + 4}{(x + \Delta x - 1)\,(x - 1)}}{\Delta x}$ Get a common denominator of $(x + \Delta x - 1)\,(x - 1)$

$= \lim\limits_{\Delta x \to 0} \dfrac{-4\Delta x}{(x + \Delta x - 1)\,(x - 1)} \cdot \dfrac{1}{\Delta x}$ Combine similar terms; invert and multiply

$= \lim\limits_{\Delta x \to 0} \dfrac{-4}{(x + \Delta x - 1)\,(x - 1)}$ Since $\Delta x \neq 0$, we can reduce using $\dfrac{\Delta x}{\Delta x} = 1$

$= \dfrac{-4}{(x - 1)\,(x - 1)}$ or $\dfrac{-4}{(x - 1)^2}$ Substitute $\Delta x = 0$ to complete the limit **(Limit Definition of Derivative)**

4. $\lim\limits_{\Delta x \to 0} \dfrac{2(-1 + \Delta x) - 3 - (2(-1) - 3)}{\Delta x}$ Replace x with $(-1 + \Delta x)$ in f

$= \lim\limits_{\Delta x \to 0} \dfrac{-2 + 2\Delta x - 3 - (-5)}{\Delta x}$ Use the distributive property

$= \lim\limits_{\Delta x \to 0} \dfrac{2\Delta x}{\Delta x} = 2$ Since $\Delta x \neq 0$, we can reduce using $\dfrac{\Delta x}{\Delta x} = 1$

Check: Using the derivative rules, $f'(x) = 2$. **(Limit Definition of Derivative)**

5. $\lim\limits_{\Delta x \to 0} \dfrac{(2 + \Delta x)^2 - 2(2 + \Delta x) + 1 - (2^2 - 2(2) + 1)}{\Delta x}$ Replace x with $(2 + \Delta x)$ in f

$= \lim\limits_{\Delta x \to 0} \dfrac{4 + 4\Delta x + (\Delta x)^2 - 4 - 2\Delta x + 1 - 1}{\Delta x}$ Expand $(2 + \Delta x)^2$ and use the distributive property

$$= \lim_{\Delta x \to 0} \frac{2\Delta x + (\Delta x)^2}{\Delta x} \quad \text{Combine similar terms}$$

$$= \lim_{\Delta x \to 0} \frac{\Delta x (2 + \Delta x)}{\Delta x} \quad \text{Factor out } \Delta x \text{ in the numerator}$$

$$= \lim_{\Delta x \to 0} (2 + \Delta x) \quad \text{Since } \Delta x \neq 0, \text{ we can reduce using } \frac{\Delta x}{\Delta x} = 1$$

$$= 2 \quad \text{Substitute } \Delta x = 0 \text{ to complete the limit}$$

Check: Using the derivative rules, $f'(x) = 2x - 2$ and $f'(2) = 2(2) - 2 = 2$. **(Limit Definition of Derivative)**

6. $\lim_{\Delta t \to 0} \dfrac{\sqrt{3(t + \Delta t) + 8} - \sqrt{3t + 8}}{\Delta t}$ Replace t with $(t + \Delta t)$ in s

$$= \lim_{\Delta t \to 0} \frac{\sqrt{3(t + \Delta t) + 8} - \sqrt{3t + 8}}{\Delta t} \cdot \frac{\sqrt{3(t + \Delta t) + 8} + \sqrt{3t + 8}}{\sqrt{3(t + \Delta t) + 8} + \sqrt{3t + 8}}$$

Multiply by the conjugate to rationalize the numerator

$$= \lim_{\Delta t \to 0} \frac{3(t + \Delta t) + 8 - (3t + 8)}{\Delta t (\sqrt{3(t + \Delta t) + 8} + \sqrt{3t + 8})}$$ Multiply the numerators, leave the denominator factored

$$= \lim_{\Delta t \to 0} \frac{3\Delta t}{\Delta t (\sqrt{3(t + \Delta t) + 8} + \sqrt{3t + 8})}$$ Combine similar terms in the numerator

$$= \lim_{\Delta t \to 0} \frac{3}{\sqrt{3(t + \Delta t) + 8} + \sqrt{3t + 8}}$$ Since $\Delta t \neq 0$, we can reduce using $\frac{\Delta t}{\Delta t} = 1$

$$= \frac{3}{\sqrt{3t + 8} + \sqrt{3t + 8}} \quad \text{Substitute } \Delta t = 0$$

$$= \frac{3}{2\sqrt{3t + 8}} \quad \text{Combine similar terms in the denominator}$$

This gives velocity at any time t. In particular, when $t = 3$,

$$\frac{3}{2\sqrt{3t + 8}} = \frac{3}{2\sqrt{3(3) + 8}} = \frac{3}{2\sqrt{17}} \quad \textbf{(Definition of Velocity)}$$

7. $f'(x) = 0$ since π is a constant and the derivative of a constant is 0. **(The Derivative)**

8. $f'(x) = 12x - 3$ **(The Derivative)**

9. Rewrite $f(x) = 2x^{-1/3} + x^{1/2}$. Then use the power rule:

$$f'(x) = \frac{-2}{3\sqrt[3]{x^4}} + \frac{1}{2\sqrt{x}} \quad \text{or} \quad \frac{-2}{3x\sqrt[3]{x}} + \frac{1}{2\sqrt{x}} \quad \textbf{(The Derivative)}$$

10. First find the slope of tangent line by finding the derivative:

$f'(x) = 6x - 2$

Then, the slope when $x = 2$ is $f'(2) = 6(2) - 2 = 10$

When $x = 2, y = f(2) = 3(2)^2 - 2(2) + 4 = 12$

Use the point-slope formula to find the equation:

$y - 12 = 10(x - 2)$ Use $y - y_1 = m(x - x_1)$

$y = 10x - 8$ Use the distributive property to remove parentheses, then add 12 to both sides of the equation to write the answer in slope-intercept form. **(Tangent Line)**

11. Use the Product Rule:

$f: 3x + 2$ $g: x^2 - 4$ List f and g

$f': 3$ $g': 2x$ Find the derivative of f and g

$\dfrac{dy}{dx} = (3x + 2)(2x) + 3(x^2 - 4)$ Use the Product Rule: $f(x) \cdot g'(x) + f'(x) \cdot g(x)$

$= 6x^2 + 4x + 3x^2 - 12$ Multiply

$= 9x^2 + 4x - 12$ Add similar terms

You can also mutiply first to write $y = 3x^3 + 2x^2 - 12x - 8$ and then $\dfrac{dy}{dx} = 9x^2 + 4x - 12$. **(Product Rule)**

12. Use the Quotient Rule:

$f: 4x + 5$ $g: 2x - 1$ List f and g where f is the numerator and g is the denominator of the given function

$f': 4$ $g': 2$ Find the derivative of f and g

$\dfrac{dy}{dx} = \dfrac{(2x - 1)(4) - (4x + 5)(2)}{(2x - 1)^2}$ Use the Quotient Rule: $\dfrac{g(x) \cdot f'(x) - f(x) \cdot g'(x)}{[g(x)]^2}$

$= \dfrac{8x - 4 - 8x - 10}{(2x - 1)^2}$ Use the distributive property to remove parentheses in the numerator

$= \dfrac{-14}{(2x - 1)^2}$ Add similar terms

$= -\dfrac{14}{(2x - 1)^2}$ **(Quotient Rule)**

13. Use the Chain Rule: $y' = 3(3x^2 + 2x)^2(6x + 2) = 6(3x + 1)(3x^2 + 2x)^2$ **(Chain Rule)**

14. Use the Product Rule:

h: $4x$ g: $(x^2 - 6x)^2$ List each factor from the product

h': 4 g': $2(x^2 - 6x)(2x - 6)$ Find the derivative of each factor

$f'(x) = 4x(2)(x^2 - 6x)(2x - 6) + 4(x^2 - 6x)^2$ Use the Product Rule: $h(x) \cdot g'(x) + h'(x) \cdot g(x)$

Now common factor $4(x^2 - 6x)$:

$f'(x) = 4(x^2 - 6x)[x(2)(2x - 6) + (x^2 - 6x)]$

$f'(x) = 4(x^2 - 6x)[4x^2 - 12x + x^2 - 6x]$ Simplify inside the brackets

$f'(x) = 4(x^2 - 6x)[5x^2 - 18x]$ Combine similar terms   **(Chain Rule)**

15. $h'(x) = 4\left(\dfrac{2x-3}{x+1}\right)^3 \left[\dfrac{(x+1)2 - (2x-3)1}{(x+1)^2}\right]$ Use the Chain Rule. Note the Quotient Rule was used to

find the derivative of $\dfrac{2x-3}{x+1}$ as part of the chain rule process.

$= 4\left(\dfrac{2x-3}{x+1}\right)^3 \dfrac{5}{(x+1)^2}$ Simplify

$= 4 \cdot \dfrac{(2x-3)^3}{(x+1)^3} \cdot \dfrac{5}{(x+1)^2}$ Raise the numerator and denominator to the third power

$= \dfrac{20(2x-3)^3}{(x+1)^5}$ Since the bases are the same, the exponents can be added. **(Chain Rule)**

16. Rewrite $g(x) = (x^2 + 4x - 1)^{1/2}$. Then:

$g'(x) = \dfrac{1}{2}(x^2 + 4x - 1)^{-1/2}(2x + 4)$ Use the Chain Rule

$= \dfrac{2(x+2)}{2(x^2 + 4x - 1)^{1/2}}$ Write the negative exponent as a positive exponent in the denominator

$= \dfrac{x+2}{\sqrt{x^2 + 4x - 1}}$ An exponent of 1/2 can be written as a square root. **(Chain Rule)**

17. $y' = (\cos(4x + 6))(4) = 4\cos(4x + 6)$ **(Derivatives of Trigonometric Functions)**

18. $y' = (-\sin(x^2))(2x) = -2x\sin(x^2)$ **(Derivatives of Trigonometric Functions)**

19. $y' = (3\sec^2 2x)(2) = 6\sec^2 2x$ **(Derivatives of Trigonometric Functions)**

20. $y' = (-\csc(2x^2-3)\cot(2x^2-3))(4x)$ Use $\dfrac{d}{dx}[\csc u] = (-\csc u \cot u)u'$

    $= -4x\csc(2x^2-3)\cot(2x^2-3)$ **(Derivatives of Trigonometric Functions)**

21. $y' = -2[\sec\pi x\tan\pi x](\pi)$ Use $\dfrac{d}{dx}[\sec u] = (\sec u\tan u)u'$

    $= -2\pi\sec\pi x\tan\pi x$ **(Derivatives of Trigonometric Functions)**

22. $y' = -4\csc^2(x+\pi/2)$ Use $\dfrac{d}{dx}[\cot u] = \left(-\csc^2 u\right)u'$ **(Derivatives of Trigonometric Functions)**

23. Use the Product Rule:

    $f: 4x^2$ $g: \cos 2x$ List the factors of the product

    $f': 8x$ $g': -2\sin 2x$ Find the derivative of each factor

    $y' = 4x^2(-2\sin 2x) + 8x(\cos 2x)$ Use the Product Rule: $f(x)\cdot g'(x) + f'(x)\cdot g(x)$

    $= -8x(x\sin 2x - \cos 2x)$ **(Derivatives of Trignometric Functions)**

24. Use the Product Rule:

    $f: \sin x$ $g: \cot x$ List the factors of the product

    $f': \cos x$ $g': -\csc^2 x$ Find the derivative of each factor

    $y' = \sin x(-\csc^2 x) + \cos x\cot x$ Use the Product Rule: $f(x)\cdot g'(x) + f'(x)\cdot g(x)$

    $= \dfrac{-\sin x}{\sin^2 x} + \cos x\dfrac{\cos x}{\sin x}$ Use the identities $\csc x = \dfrac{1}{\sin x}$ and $\cot x = \dfrac{\cos x}{\sin x}$

    $= -\dfrac{1}{\sin x} + \dfrac{\cos^2 x}{\sin x}$ Reduce the fraction

    $= \dfrac{-1+\cos^2 x}{\sin x}$ Add the fractions

    $= -\dfrac{\sin^2 x}{\sin x}$ Use the identity $-1+\cos^2 x = -\sin^2 x$

    $= -\sin x$ Reduce the fraction

Note that this answer could be written in several forms. The student should be prepared to use trignometric identities to manipulate the answer as necessary. Also, you may recognize that this problem could be more easily handled by using trigonometric identities *before* taking the derivative:

    $\sin x\cot x = \sin x\cdot\dfrac{\cos x}{\sin x} = \cos x$, and now the derivative can be seen immediately as $-\sin x$. **(Derivatives of Trignometric Functions)**

25. Use the Quotient Rule:

f: $\sin x$ g: $1 + \tan x$ List the numerator and denominator

f': $\cos x$ g': $\sec^2 x$ Find the derivative of the numerator and denominator

$y' = \dfrac{(1 + \tan x)\cos x - \sin x \sec^2 x}{(1 + \tan x)^2}$ Use the Quotient Rule: $\dfrac{g(x) \cdot f'(x) - f(x) \cdot g'(x)}{[g(x)]^2}$

$= \dfrac{\cos x + \sin x - \sin x \sec^2 x}{(1 + \tan x)^2}$ Use $\tan x \cdot \cos x = \dfrac{\sin x}{\cos x} \cdot \cos x = \sin x$

(Derivatives of Trignometric Functions)

26. It may help to write $\sin^2 2x$ as $(\sin 2x)^2$. Then

$y' = 2(\sin 2x)^1 (\cos 2x)(2)$ Use the Chain Rule

$= 4\sin 2x \cos 2x$

This answer may appear in several forms. For example, using the identity $\sin 2\theta = 2\sin\theta\cos\theta$, we could write:

$y' = 2(2\sin 2x \cos 2x)$ Factor out 2

$= 2(\sin 2(2x))$ Use the double angle identity $\sin 2\theta = 2\sin\theta\cos\theta$

$= 2\sin 4x$ **(Derivatives of Trignometric Functions)**

27. $y = \tan^3 (3x + 1) = [\tan(3x + 1)]^3$ Rewrite with the exponent outside brackets

$y' = 3[\tan(3x + 1)]^2 \sec^2 (3x + 1)(3)$ Use the Chain Rule

$= 9\tan^2 (3x + 1) \sec^2 (3x + 1)$ **(Derivatives of Trignometric Functions)**

28. Use the Product Rule:

f: $\csc 2x$ g: $\sec 3x$ List each factor

f': $-2\csc 2x \cot 2x$ g': $3\sec 3x \tan 3x$ Find the derivative of each factor

$y' = \csc 2x (3\sec 3x \tan 3x) - 2\csc 2x \cot 2x (\sec 3x)$ Use the Product Rule: $f(x) \cdot g'(x) + f'(x) \cdot g(x)$

(Derivatives of Trignometric Functions)

29. $y' = 4\cos^3 (3x + 1)(-\sin(3x + 1))(3)$ Use the Chain Rule

$= -12\cos^3 (3x + 1)\sin(3x + 1)$ **(Derivatives of Trignometric Functions)**

30. Write $y = \sqrt{\sin x}$ as $y = (\sin x)^{1/2}$. Then $y' = \dfrac{1}{2}(\sin x)^{-1/2}\cos x$ or $y' = \dfrac{\cos x}{2\sqrt{\sin x}}$ **(Derivatives of Trig-onometric Functions)**

31. $f{:}\ x \qquad g{:}\ \sin x \qquad$ List each factor

$f'{:}\ 1 \qquad g'{:}\ \cos x \qquad$ Find the derivative of each factor

$\dfrac{dy}{dx} = x\cos x + \sin x \qquad$ Use the Product Rule: $f(x) \cdot g'(x) + f'(x) \cdot g(x)$

To find the second derivative, use the Product Rule to find the derivative of $x\cos x$:

$h{:}\ x \qquad r{:}\ \cos x \qquad$ List each factor

$h'{:}\ 1 \qquad r'{:}\ -\sin x \qquad$ Find the derivative of each factor

$\dfrac{d^2y}{dx^2} = -x\sin x + \cos x + \cos x \qquad$ Use the Product Rule, and the derivative of a sum is
the sum of the derivatives.

$\qquad = -x\sin x + 2\cos x \qquad$ Add similar terms $\qquad$ **(Higher-Order Derivatives)**

32. $\dfrac{dy}{dx} = 2\cos x\,(-\sin x) = -2\sin x \cos x = -\sin 2x$

$\dfrac{d^2y}{dx^2} = -\cos 2x\,(2) = -2\cos 2x$ **(Higher-Order Derivatives)**

33. Although the student could take 50 derivatives, it would be better to look for a pattern:

$\dfrac{dy}{dx} = \cos x \qquad$ Find the first derivative

$\dfrac{d^2y}{dx^2} = -\sin x \qquad$ Find the second derivative

$\dfrac{d^3y}{dx^3} = -\cos x \qquad$ Find the third derivative

$\dfrac{d^4y}{dx^4} = \sin x \qquad$ Find the fourth derivative

and the pattern begins again, repeating in blocks of 4. Since $50 \div 4 = 12$ with a remainder of 2, then $\dfrac{d^{50}y}{dx^{50}}$

equals $\dfrac{d^2y}{dx^2}$ or $-\sin x$. **(Higher-Order Derivatives)**

34. This notation means find the second derivative (not the first derivative squared).

$D_x(\tan 2x) = 2\sec^2 2x \qquad$ Use $\dfrac{d}{dx}[\tan u] = \left(\sec^2 u\right) u'$

$D_x^2(\tan 2x) = D_x(2\sec^2 2x) \qquad$ To find the second derivative, take the derivative of the first derivative

$$= 4\,(\sec 2x)\,(\sec 2x \tan 2x)\,2 \qquad \text{Use the Chain Rule}$$

$$= 8 \sec^2 2x \tan 2x$$

35. $y' = 18x^2 + 16x - 5$ Find the first derivative

    $y'' = 36x + 16$ Find the derivative of the first derivative

    $y''' = 36$ Find the derivative of the second derivative   **(Higher-Order Derivatives)**

36. $y' = 4\,(2x+5)^3\,(2)$

    $y' = 8\,(2x+5)^3$ Simplify before taking the next derivative

    $y'' = 24\,(2x+5)^2\,(2)$ Find the second derivative

    $y'' = 48\,(2x+5)^2$ Simplify   **(Higher-Order Derivatives)**

37. Use implicit differentiation. Note that the Product Rule is used to find the derivative of $2xy$ as follows:

    f: $2x$ g: y List each factor

    f': 2 g': y' Find each derivative.

    The Product Rule gives: $2xy' + 2y$. Now put this into the derivative of the given equation:

    $2x + 2xy' + 2y + 2yy' = 0$

    $y'(2x + 2y) = -2x - 2y$ Get all terms containing y' on one side of the equation, all other terms on the other side. Factor y' out.

    $y' = \dfrac{-2x - 2y}{2x + 2y} = -1$ Solve for y' **(Implicit Differentiation)**

38. Use the Product Rule on $x^2 y^2$:

    f: x^2 g: y^2 List each factor

    f': $2x$ g': $2yy'$ Find the derivative of each factor

    $x^2 2yy' + 2xy^2 - 3 = 0$

    $x^2 2yy' = -2xy^2 + 3$ Isolate the term containing y'

    $y' = \dfrac{-2xy^2 + 3}{2x^2 y}$ Divide both sides of the equation by $2x^2 y$ to solve for y' **(Implicit Differentiation)**

39. Use implicit differentiation:

    $2x + (\cos y)\,y' = 0$ Use implicit differentiation on $\sin y$

       $(\cos y)\,y' = -2x$ Isolate the term containing y'

$$y' = \frac{-2x}{\cos y}$$ Divide both sides of the equation by cos y to solve for y'

(Implicit Differentiation)

40. Use implicit differentiation:

$$(-\sin xy)\,(x\,y'+y) = 8x$$ Use the Product Rule to find the derivative of xy

$$-x\,y'\sin xy - y\sin xy = 8x$$ Multiply $-\sin xy$ times xy' and y

$$-xy'\sin xy = 8x + y\sin xy$$ Isolate the term containing y'

$$y' = -\frac{8x + y\sin xy}{x\sin xy}$$ Divide both sides of the equation by $x\sin xy$

(Implicit Differentiation)

41. Use implicit differentiation:

$$y\sec^2 x + y'\tan x - 2x = 0$$ Use the Product Rule to find the derivative of $y\tan x$

$$y'\tan x = 2x - y\sec^2 x$$ Isolate the term containing y'

$$y' = \frac{2x - y\sec^2 x}{\tan x}$$ Divide both sides of the equation by $\tan x$

(Implicit Differentiation)

42. Find the slope of the tangent line by finding y':

$$8x + 2y\,y' = 0$$ Use implicit differentiation

$$y' = \frac{-8x}{2y} = \frac{-4x}{y}$$ Isolate the term containing y', then solve for y' by dividing both sides by $2y$

At $(3, 8)$, $y' = \dfrac{-4\,(3)}{8} = \dfrac{-3}{2}$.   Then the equation of the tangent line is $y - 8 = -\dfrac{3}{2}\,(x - 3)$ or

$$y = -\frac{3}{2}x + \frac{25}{2}$$ **(Implicit Differentiation)**

43. Find the slope of the tangent line by finding y':

$$2x + 8y\,y' - 2 = 0$$

$$y' = \frac{2 - 2x}{8y} = \frac{1 - x}{4y}\,.$$

The slope at $\left(2, \dfrac{\sqrt{3}}{2}\right)$ is $y' = \dfrac{1 - 2}{4\left(\dfrac{\sqrt{3}}{2}\right)} = \dfrac{-1}{2\sqrt{3}}$. The equation of the tangent line is $y - \dfrac{\sqrt{3}}{2} = -\dfrac{1}{2\sqrt{3}}\,(x - 2)$

(Implicit Differentiation)

Grade Yourself

Circle the question numbers that you had incorrect. Then indicate the number of questions you missed. If you answered more than three questions incorrectly, you will then have to focus on that topic. If a topic has less than three questions and you had at least one wrong, we suggest you study that topic also. Read your textbook, a review book, or ask your teacher for help.

Subject: Derivatives

Topic	Question Numbers	Number Incorrect
Limit Definition of Derivative	1, 2, 3, 4, 5	
Definition of Velocity	6	
The Derivative	7, 8, 9	
Tangent Line	10	
Product Rule	11	
Quotient Rule	12	
Chain Rule	13, 14, 15, 16	
Derivatives of Trigonometric Functions	17, 18, 19, 20, 21, 22, 23, 24, 25, 26, 27, 28, 29, 30	
Higher-Order Derivatives	31, 32, 33, 34, 35, 36	
Implicit Differentiation	37, 38, 39, 40, 41, 42, 43	

Applications of Derivatives

4

 Brief Yourself

After a student has gained some mastery of differentiation, he/she will be expected to make use of those skills to solve different types of application problems. This test chapter includes related rates problems, graphing problems, and max-min problems.

It is usually helpful to draw and label a sketch that reflects the information given in the problem. Also, note the units involved in the problems. An area must be measured in square units (such as ft^2, m^2), a volume in cubic units (such as $in.^3$, m^3). When you encounter units such as ft/sec, these are rates of change with respect to time, and represent derivatives. Label your answers appropriately. Reread the given problem to make sure you have answered the question(s).

You may need to review some geometric formulas for area and volume. Some of those formulas are listed here:

Area of a square: $A = e^2$ Volume of a cube: $V = e^3$, where e is the length of one side

Circumference of a circle: $C = \pi d$, where d is the length of the diameter   Area of a circle: $A = \pi r^2$

Volume of a sphere: $V = \frac{4}{3}\pi r^3$, where r is the radius

Volume of a right circular cylinder: $V = \pi r^2 h$, where r is the radius and h is the height

Volume of a right circular cone: $V = \frac{1}{3}\pi r^2 h$, where r is the radius and h is the height

Since the student must be able to find a derivative implicitly with respect to time to solve related rates problems, it might be wise to review the topic of implicit differentiation.

Some of the questions about graphs require the use of the first derivative, and others also require the use of the second derivative. Questions about intervals where the graph is increasing or decreasing require the use of the first derivative. Questions about the concavity of a graph or inflection points require the use of the second derivative. When the second derivative is positive, the graph is concave up. When the second derivative is negative, the graph is concave down. To find relative maximums or minimums, you may use the first derivative to find the critical value(s) and test to the left and right of each value in the first derivative (the First Derivative Test). Or, you may use the Second Derivative Test to test critical values in the second derivative.

A graphing calculator can be a valuable tool for finding and/or checking answers. Adjust the viewing screen as appropriate. Keep in mind that an increasing interval contains a portion of a graph that is rising from left to right, and a decreasing interval contains a portion of a graph that is falling from left to right. A graph is concave up where it appears able to hold water (tangent lines drawn to the curve will be below the curve), and concave down where it would dump water (tangent lines drawn to the curve will be above the curve). An absolute maximum occurs at the highest y value on the graph. You are guaranteed an absolute maximum and an absolute minimum only when the function is continuous and defined on a closed interval. Otherwise, there may be an absolute maximum and/or absolute minimum or neither.

Remember that the Second Derivative test fails if the second derivative of a critical value equals 0. In that case, you must return to the First Derivative Test to determine whether the critical value yields a maximum, minimum, or neither.

Test Yourself

1. Find the rate of change of the volume of a cube with respect to time.

2. Find the rate of change of the area of a square with respect to time.

3. Find the rate of change of the volume of a right circular cone with respect to time.

4. If the length of the edge of a cube is increasing at a rate of 5 cm/sec, find the rate of change of the volume of the cube when an edge is 12 cm.

5. If the area of a square is decreasing at a rate of $50 \text{ ft}^2/\text{sec}$, find the rate of change of the length of a side when a side is 18 ft long. Is the length of the side increasing or decreasing?

6. If the radius of a right circular cone is a constant 10 cm and the height is increasing at a rate of 6 cm/min, how fast is the volume of the cone increasing when the height is 4.5 cm?

7. Oil spills into a lake in a circular pattern. If the radius of the circle increases at a rate of 1 foot per minute, how fast is the area of the spill increasing at the end of 1 hour?

8. Air is being pumped into a spherical balloon at a rate of 16 cubic inches per second. Find the rate of change of the radius when the radius is 8.5 inches.

9. A hot-air balloon lifts off the ground 75 feet from an observer. If the balloon rises at a rate of 10 ft/min, how fast is the angle of elevation changing when the balloon is 100 feet high?

10. A 10-foot ladder is leaning against the wall of a house. The base of the ladder slides away from the wall at a rate of 2 in/sec. Find the rate at which the top of the ladder slides down the wall when the base is 3 feet from the wall.

11. Find dy for $y = 7x^2 - 8x + 1$.

12. Find dy for $y = (6x - 1)^3$.

13. Find dy for $y = 2x \sin x$.

14. Use differentials to approximate $\sqrt{16.2}$.

15. Use differentials to approximate $\sqrt[3]{26.8}$.

16. Find the intervals on which $f(x) = 2x^2 + 6x$ is increasing or decreasing.

17. Find the intervals on which
$$f(x) = x^3 + 4x^2 + 4$$
is increasing or decreasing.

18. Find the intervals on which $f(x) = \sqrt{2x - 3}$ is increasing or decreasing.

19. Find the intervals on which $f(x) = \dfrac{x^2}{x^2 - 9}$ is increasing or decreasing.

20. Find the intervals where $f(x) = \sin x + 1$ is increasing or decreasing on $[0, 2\pi]$.

21. Find the intervals where $f(x) = \tan x$ is increasing or decreasing.

22. Find the absolute maximum and minimum for $f(x) = \cos x$ on $[0, 2\pi]$.

23. Find the absolute maximum and minimum for $f(x) = -x^2 + 4x$ on $[-1, 3]$.

24. Find the absolute maximum and minimum for $f(x) = \sqrt{4 - x}$ on $[0, 4]$.

25. Find all relative extrema for $f(x) = 2x^2 + 6x$.

26. Find all relative extrema for $f(x) = 2x^3 - 3x^2 - 36x + 4$.

27. Find all relative extrema for $f(x) = \sin x + \cos x$ on $[0, 2\pi]$.

28. Determine the intervals where $f(x) = x^3 - x$ is concave upward or downward.

29. Determine the intervals where $f(x) = x^3 - x^2 - x - 2$ is concave upward or downward.

30. Determine the intervals where $f(x) = x^4 - 5x^2 + 6$ is concave upward or downward.

31. Determine the intervals where $f(x) = \dfrac{4}{x^2 - 9}$ is concave upward or downward.

32. Find all inflection points for $f(x) = x^3 - x$.

33. Find all inflection points for $f(x) = x^4 - 5x^2 + 6$.

34. Find all inflection points for $f(x) = 4x^2 - 3x + 2$.

35. Find all inflection points for $f(x) = \dfrac{4}{x^2 - 9}$.

36. Use the Second Derivative Test to find the relative extrema or state that the test does not apply: $f(x) = x^3 - 4x^2 + 1$

37. Use the Second Derivative Test to find the relative extrema or state that the test does not apply: $f(x) = x^4 - 5x^2 + 4$

38. Use the Second Derivative Test to find the relative extrema or state that the test does not apply: $f(x) = x^3 - x^2 - x - 2$

39. Find $\displaystyle \lim_{x \to 2^+} \dfrac{6x}{x^2 - 4}$.

40. Find $\displaystyle \lim_{x \to 2^-} \dfrac{6x}{x^2 - 4}$.

41. Find $\displaystyle \lim_{x \to -2^+} \dfrac{6x}{x^2 - 4}$.

42. Find $\displaystyle \lim_{x \to -1^-} \dfrac{2}{x + 1}$.

43. Find $\displaystyle \lim_{x \to 3^+} \dfrac{1}{(x - 3)^2}$.

44. Find each of the following and then use all the information to prepare a graph of $f(x) = x^2 + 2x + 1$:

    a) domain

    b) x- and y-intercepts

    c) symmetry with respect to the y-axis and origin

    d) horizontal and vertical asymptotes

    e) relative extrema

    f) intervals where f is increasing or decreasing

    g) intervals where f is concave upward or downward

    h) inflection points

45. Find each of the following and then use all the information to prepare a graph of

$$f(x) = \frac{6x}{x^2 - 4}:$$

a) domain
b) x- and y-intercepts
c) symmetry with respect to the y-axis and origin
d) horizontal and vertical asymptotes
e) relative extrema
f) intervals where f is increasing or decreasing
g) intervals where f is concave upward or downward
h) inflection points

46. Find each of the following and then use all the information to prepare a graph of

$$f(x) = (x + 1)^2 (x - 2):$$

a) domain

b) x- and y-intercepts
c) symmetry with respect to the y-axis and origin
d) horizontal and vertical asymptotes
e) relative extrema
f) intervals where f is increasing or decreasing
g) intervals where f is concave upward or downward
h) inflection points

47. Find two positive numbers whose sum is 48 and whose product is a maximum.

48. Find two positive numbers whose product is 48 and whose sum is a minimum. Find the minimum.

49. A gardener has 600 square feet available to enclose two adjacent areas for gardens. The gardens are to be identical rectangular shapes enclosing the entire 600 square feet. What dimensions should be used to minimize the amount of wire fencing required?

Check Yourself

1. Let e = the length of an edge. Then,

$$V = e^3 \qquad \text{Use the formula for the volume of a cube}$$

$$\frac{dV}{dt} = 3e^2\frac{de}{dt} \qquad \text{Take the derivative with respect to time of each side of the equation} \quad \textbf{(Related Rates)}$$

2. Let s = the length of a side. Then,

$$A = s^2 \qquad \text{Use the formula for the area of a square}$$

$$\frac{dA}{dt} = 2s\frac{ds}{dt} \qquad \text{Take the derivative with respect to time of each side of the equation} \quad \textbf{(Related Rates)}$$

3. $V = \frac{1}{3}\pi r^2 h \qquad$ Use the formula for the volume of a right circular cone

$$\frac{dV}{dt} = \frac{1}{3}\pi\left[r^2\frac{dh}{dt} + 2r\frac{dr}{dt}h\right] \qquad \text{Take the derivative of each side of the equation with respect to time}$$

(Related Rates)

4. Given $\dfrac{de}{dt} = 5$ cm/sec and $e = 12$ cm

$\dfrac{dV}{dt} = 3e^2\dfrac{de}{dt}$ Use the derivative found in problem 1. Do *not* substitute $e = 12$ into the formula for V

before computing $\dfrac{dV}{dt}$.

$\dfrac{dV}{dt} = 3\,(12 \text{ cm})^2\,(5 \text{ cm/sec})$ Substitute the given information into the derivative

$= 3\,(144 \text{ cm}^2)\,(5 \text{ cm/sec})$ Square 12 cm

$= 2160 \text{ cm}^3/\text{sec}$ Multiply

Note that the units match what is requested—that is, a rate of change of volume (cubic units) with respect to time (seconds, in this problem). **(Related Rates)**

5. Given $\dfrac{dA}{dt} = -50$ ft²/sec , where the negative sign represents the fact that the area is decreasing, and $s = 18$ ft.

$\dfrac{dA}{dt} = 2s\dfrac{ds}{dt}$ Use the derivative found in problem 2

$-50 \text{ ft}^2/\text{sec} = 2\,(18 \text{ ft})\dfrac{ds}{dt}$ Substitute the given information into the derivative

$\dfrac{-50 \text{ ft}^2/\text{sec}}{36 \text{ ft}} = \dfrac{ds}{dt}$ Divide both sides of the equation by 36 feet

$-\dfrac{25}{18}$ ft/sec $= \dfrac{ds}{dt}$ Reduce the fraction

Note that the units can be treated like any other fraction: $\dfrac{\text{ft}^2}{\text{sec}} \div \text{ft} = \dfrac{\text{ft}^2}{\text{sec}} \cdot \dfrac{1}{\text{ft}} = \dfrac{\text{ft}}{\text{sec}}$. The negative sign indicates that the side length is decreasing (of course, it must be decreasing since the area is decreasing). **(Related Rates)**

6. $V = \pi r^2 h$ Use the formula for the volume of a right circular cone

$\dfrac{dV}{dt} = \pi\left[r^2\dfrac{dh}{dt} + 2r\dfrac{dr}{dt}h\right]$ Take the derivative with respect to time of both sides of the equation (see solution to problem #3).

Given $r = 10$ cm, $\dfrac{dh}{dt} = 6$ cm/min, $h = 4.5$ cm, and $\dfrac{dr}{dt} = 0$ (since the radius length is not changing). Then:

$\dfrac{dV}{dt} = \pi\,[\,(10 \text{ cm})^2\,(6 \text{ cm/min}) + 2\,(10 \text{ cm})\,(0)\,(4.5 \text{ cm})\,]$ Substitute the given information into the derivative

$\dfrac{dV}{dt} = 600\pi$ cm^3/min   Simplify the right side   **(Related Rates)**

7. $A = \pi r^2$ Use the formula for the area of a circle

$\dfrac{dA}{dt} = 2\pi r \dfrac{dr}{dt}$ Take the derivative with respect to time of both sides of the equation

Given $\dfrac{dr}{dt} = 1$ foot/min, at the end of 1 hour = 60 min, the radius is 60 feet. Then:

$\dfrac{dA}{dt} = 2\pi\,(60\text{ feet})(1\text{ foot/min})$ Substitute the given information into the derivative

$\qquad = 120\pi$ ft^2/min or approximately 376.99 ft^2/min   **(Related Rates)**

8. $V = \dfrac{4}{3}\pi r^3$ Use the formula for the volume of a sphere

$\dfrac{dV}{dt} = 4\pi r^2 \dfrac{dr}{dt}$ Take the derivative with respect to time of both sides of the equation

16 in^3/sec $= 4\pi\,(8.5\text{ in})^2 \dfrac{dr}{dt}$ Substitute the given information into the derivative

$\dfrac{16\text{ in}^3/\text{sec}}{4\pi\,(8.5\text{ in})^2} = \dfrac{dr}{dt}$ Solve for $\dfrac{dr}{dt}$

0.0176 in/sec $= \dfrac{dr}{dt}$ **(Related Rates)**

9. Draw and label a sketch. Let θ = angle of elevation. Given $\dfrac{dy}{dt} = 10$ ft/min.

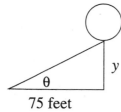

75 feet

$\tan\theta = \dfrac{y}{75}$ Use the definition of tangent

$\sec^2\theta \dfrac{d\theta}{dt} = \dfrac{1}{75}\dfrac{dy}{dt}$ Take the derivative with respect to time

When $y = 100$,

$\tan\theta = \dfrac{100}{75} = \dfrac{4}{3}$

Since $1 + \tan^2\theta = \sec^2\theta$,

$$1 + \left(\frac{4}{3}\right)^2 = \sec^2\theta$$

$$1 + \frac{16}{9} = \sec^2\theta$$

$$\frac{25}{5} = \sec^2\theta$$

Now substitute into $\sec^2\theta\dfrac{d\theta}{dt} = \dfrac{1}{75}\dfrac{dy}{dt}$:

$$\frac{25}{9}\frac{d\theta}{dt} = \frac{1}{75}(10 \text{ ft/min})$$

$$\frac{d\theta}{dt} = \left(\frac{9}{25}\right)\left(\frac{1}{75}\right)(10 \text{ ft/min})$$

$$\frac{d\theta}{dt} = \frac{6}{125} \approx 0.05 \text{ rad/min}$$

(Related Rates)

10.

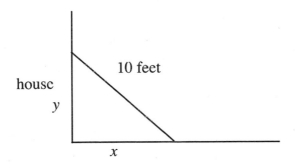

Using the Pythagorean Theorem,

$$10^2 = x^2 + y^2 \qquad \text{Use } c^2 = a^2 + b^2$$

$$0 = 2x\frac{dx}{dt} + 2y\frac{dy}{dt} \qquad \text{Take the derivative with respect to time of both sides of the equation}$$

$$\frac{dy}{dt} = \frac{-x\,dx}{y\,dt} \qquad \text{Solve for } \frac{dy}{dt}$$

When $x = 3$ feet, we have

$$10^2 = 3^2 + y^2 \qquad \text{Use } c^2 = a^2 + b^2$$

$$91 = y^2 \qquad \text{Square and then subtract 9 from both sides}$$

$$y = \sqrt{91} \approx 9.54 \text{ feet} \qquad \text{Take the square root of both sides of the equation}$$

Then $\dfrac{dy}{dt} = \dfrac{-(3 \text{ feet})}{\sqrt{91} \text{ feet}}$ (2 inches/second) ≈ -0.63 in/sec   **(Related Rates)**

11. $dy = (14x - 8) \, dx$ **(Differentials)**

12. $dy = 3(6x - 1)^2 (6) \, dx = 18(6x - 1)^2 dx$ **(Differentials)**

13. $dy = (2x\cos x + 2\sin x) \, dx$ **(Differentials)**

14. Let $x = 16$, $\Delta x = 0.2$, $f(x) = \sqrt{x}$. Use $f(x + \Delta x) \approx f(x) + f'(x)\Delta x$:

$f(x) = \sqrt{x} = x^{1/2}$ so

$f'(x) = \dfrac{1}{2}x^{-1/2} = \dfrac{1}{2\sqrt{x}}$ Find the derivative

$\sqrt{16.2} \approx \sqrt{16} + \dfrac{1}{2\sqrt{16}} (0.2) = 4.025$ Substitute the data into the formula   **(Differentials)**

15. Let $x = 27$, $\Delta x = -0.2$, $f(x) = \sqrt[3]{x}$.

Then $f'(x) = \dfrac{1}{3}x^{-2/3} = \dfrac{1}{3\sqrt[3]{x^2}}$ Find the derivative

$\sqrt[3]{26.8} \approx \sqrt[3]{27} + \dfrac{1}{3\sqrt[3]{27^2}} (-0.2) = 2.99$ **(Differentials)**

16.

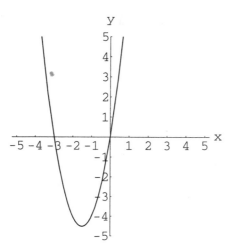

$f'(x) = 4x + 6$ Find the first derivative

Find critical values by setting $f'(x) = 0$:

$4x + 6 = 0$ Set the first derivative equal to 0

$x = -\dfrac{3}{2}$ Solve for x

Test the sign of $f'(x)$ on both sides of $-\dfrac{3}{2}$; for example,

$$f'(-2) = 4(-2) + 6 = -2 \text{ and } f'(0) = 4(0) + 6 = 6$$

Where $f'(x) > 0$, f is increasing and where $f'(x) < 0$, f is decreasing. So $f(x)$ is increasing on $\left(-\dfrac{3}{2}, \infty\right)$ and decreasing on $\left(-\infty, -\dfrac{3}{2}\right)$. **(Increasing and Decreasing Functions)**

17. By sketching the graph, we can observe generally where the graph increases and decreases.

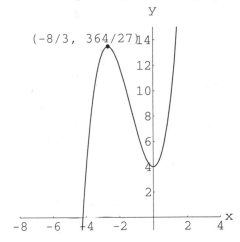

However, by using the first derivative to find critical points, and testing the sign of the first derivative in each of the regions these critical points form, we can tell exactly where the graph of f increases and decreases:

$f'(x) = 3x^2 + 8x$ Find the first derivative

$3x^2 + 8x = 0$ Set the first derivative equal to 0

$x(3x + 8) = 0$ Factor the left side

$x = 0, \quad x = -\dfrac{8}{3}$ Set each factor equal to 0 and solve for x

$f'(x)$ +   −   +

−8/3   0

$f(x)$ is increasing on $\left(-\infty, -\dfrac{8}{3}\right)$ and $(0, \infty)$. $f(x)$ is decreasing on $\left(-\dfrac{8}{3}, 0\right)$. **(Increasing and Decreasing Functions)**

18. $f(x) = \sqrt{2x-3} = (2x-3)^{1/2}$ Write the function with a fractional exponent

$f'(x) = \dfrac{1}{2}(2x-3)^{-1/2}(2) = \dfrac{1}{\sqrt{2x-3}}$ Find the first derivative

The only critical value occurs where the denominator equals 0 $\left(2x - 3 = 0 \text{ when } x = \dfrac{3}{2}\right)$. Since the domain of f is $[\frac{3}{2}, \infty)$, you only need to check to the right of $x = \frac{3}{2}$. When $x > \frac{3}{2}$, $f'(x) > 0 \Rightarrow f(x)$ is increasing on $\left(\frac{3}{2}, \infty\right)$. **(Increasing and Decreasing Functions)**

19.

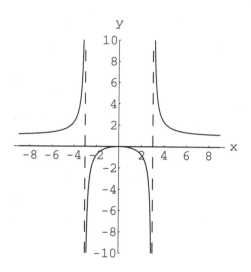

Find $f'(x)$ using the Quotient Rule: $f'(x) = \dfrac{(x^2 - 9)(2x) - x^2(2x)}{(x^2 - 9)^2} = \dfrac{-18x}{(x^2 - 9)^2}$

Critical Values:

$-18x = 0$ or $x^2 - 9 = 0$ Set the numerator equal to 0; set the denominator equal to 0

$x = 0$ or $x = \pm 3$ Solve each equation for x

Set up a number line with the cricial points. Check the sign of the derivative around the critical points:

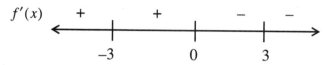

$f(x)$ is increasing on $(-\infty, -3)$ and $(-3, 0)$. $f(x)$ is decreasing on $(0, 3)$ and $(3, \infty)$. **(Increasing and Decreasing Functions)**

20. $f(x) = \sin x + 1$

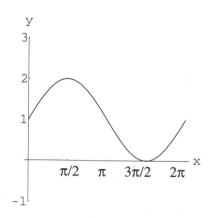

$f'(x) = \cos x$ and $\cos x = 0$ when $x = \dfrac{\pi}{2}, \dfrac{3\pi}{2}$

$f(x)$ is increasing on $\left(0, \dfrac{\pi}{2}\right)$ and $\left(\dfrac{3\pi}{2}, 2\pi\right)$. $f(x)$ is decreasing on $\left(\dfrac{\pi}{2}, \dfrac{3\pi}{2}\right)$.

(Increasing and Decreasing Functions)

21. $f'(x) = \sec^2 x$ Since $\sec x \neq 0$, and $\sec x$ is undefined when $x = \dfrac{\pi}{2}, \dfrac{3\pi}{2}, \dots$, the only critical values occur at

$\dfrac{(2n+1)\pi}{2}$, for n an integer. Since $\sec^2 x$ is always positive, $f(x)$ increases on its domain. **(Increasing and Decreasing Functions)**

22. $f'(x) = -\sin x$; $\quad -\sin x = 0$ when $x = 0, \pi, 2\pi$.

x	$f(x)$
0	1
π	−1
2π	1

Thus, the absolute maximum is 1 and the absolute minimum is −1. **(Absolute Extrema)**

23. $f'(x) = -2x + 4$ Find the first derivative

$-2x + 4 = 0$ Set the first derivative equal to 0 to find the critical point

$x = 2$ Solve for x

Find the y-values for the endpoints of the closed interval and for the critical value:

$f(-1) = -5 \quad f(2) = 4 \quad f(3) = 3$

The absolute minimum is −5, the absolute maximum is 4. **(Absolute Extrema)**

24. $f'(x) = \dfrac{1}{2}(4-x)^{-1/2}(-1) = -\dfrac{1}{2\sqrt{4-x}}$. The critical value is $x = 4$. $f(0) = 2, f(4) = 0$. The absolute minimum is 0, the absolute maximum is 2. **(Absolute Extrema)**

25. $f'(x) = 4x + 6$, $f'(x) = 0$ when $x = -\dfrac{3}{2}$. When $x < -\dfrac{3}{2}$, $f'(x) < 0$. When $x > -\dfrac{3}{2}$, $f'(x) > 0$. Therefore, $\left(-\dfrac{3}{2}, -\dfrac{9}{2}\right)$ is a relative minimum. Or note $f''\left(-\dfrac{3}{2}\right) = 4 > 0$ which implies that $\left(-\dfrac{3}{2}, -\dfrac{9}{2}\right)$ is a relative minimum by the second derivative test. **(Relative Extrema)**

26. $f'(x) = 6x^2 - 6x - 36$ Find the first derivative

    $6x^2 - 6x - 36 = 0$ Set the derivative equal to 0

    $6(x^2 - x - 6) = 0$ Factor

    $6(x - 3)(x + 2) = 0$ Factor

    Set each factor equal to 0, solve and find the critical values are $x = 3$ and -2.

$$f'(x) \quad + \quad - \quad +$$

$$-2 \qquad 3$$

    $(-2, 48)$ is a relative maximum. $(3, -77)$ is a relative minimum. **(Relative Extrema)**

27. $f'(x) = \cos x - \sin x$ Find the first derivative

    $\cos x - \sin x = 0$ Set the derivative equal to 0

    $\cos x = \sin x$ Add sin x to both sides of the equation

    $1 = \tan x$ Divide both sides of the equation by cos x

    $x = \dfrac{\pi}{4}, \dfrac{5\pi}{4}$ which are special angles. Tangent x is positive in quadrants I and III.

$$f'(x) \quad + \quad - \quad +$$

$$0 \qquad \pi/4 \qquad 5\pi/4 \qquad 2\pi$$

    The signs for $f'(x)$ can be found usin *any* anygle in the interval, but 0, π, and 2π are convenient values to substitute into $f'(x)$.

    $f\left(\dfrac{\pi}{4}\right) = \sqrt{2}$ is a relative maximum. $f\left(\dfrac{5\pi}{4}\right) = -\sqrt{2}$ is a relative minimum. **(Relative Extrema)**

28. We need the second derivative to find concavity:

    $f'(x) = 3x^2 - 1$ Find the first derivative

    $f''(x) = 6x$ Find the second derivative

    $6x = 0$ Set the second derivative equal to 0

    $x = 0$ Divide both sides of the equation by 6

    Test on both sides of 0 to determine when the second derivative is positive and negative.

When $x < 0$, $f''(x) < 0$. Therefore, $f(x)$ is concave downward on $(-\infty, 0)$

When $x > 0$, $f''(x) > 0$. Therefore, $f(x)$ is concave upward on $(0, \infty)$ **(Concavity)**

29. $f'(x) = 3x^2 - 2x - 1$ Find the first derivative

     $f''(x) = 6x - 2$ Find the second derivative

     $6x - 2 = 0$ Set the second derivative equal to 0

     $x = \dfrac{1}{3}$ Solve for x

When $x < \dfrac{1}{3}$, $f''(x) < 0$. Therefore, $f(x)$ is concave downward on $\left(-\infty, \dfrac{1}{3}\right)$

When $x > \dfrac{1}{3}$, $f''(x) > 0$. Therefore, $f(x)$ is concave upward on $\left(\dfrac{1}{3}, \infty\right)$ **(Concavity)**

30. $f'(x) = 4x^3 - 10x$ Find the first derivative

     $f''(x) = 12x^2 - 10$ Find the second derivative

     $12x^2 - 10 = 0$ Set the second derivative equal to 0

     $x^2 = \dfrac{10}{12}$ Add 10 to both sides of the equation; divide both sides of the equation by 12

     $x = \pm\sqrt{\dfrac{5}{6}} = \pm\dfrac{\sqrt{30}}{6}$ Take the square root of both sides of the equation; simplify the radical

When $x < -\dfrac{\sqrt{30}}{6}$, $f''(x) > 0$ Therefore $f(x)$ is concave upward on $\left(-\infty, -\dfrac{\sqrt{30}}{6}\right)$

When $-\dfrac{\sqrt{30}}{6} < x < \dfrac{\sqrt{30}}{6}$, $f''(x) < 0$ Therefore $f(x)$ is concave downward on $\left(-\dfrac{\sqrt{30}}{6}, \dfrac{\sqrt{30}}{6}\right)$

When $x > \dfrac{\sqrt{30}}{6}$, $f''(x) > 0$ Therefore $f(x)$ is concave upward on $\left(\dfrac{\sqrt{30}}{6}, \infty\right)$

The final answer is: $f(x)$ is concave upward on $\left(-\infty, -\dfrac{\sqrt{30}}{6}\right)$ and $\left(\dfrac{\sqrt{30}}{6}, \infty\right)$. $f(x)$ is concave downward on

$\left(-\dfrac{\sqrt{30}}{6}, \dfrac{\sqrt{30}}{6}\right)$. **(Concavity)**

31. Write the function as $f(x) = 4\left(x^2 - 9\right)^{-1}$ to use the Chain Rule to find the first derivative

     $f'(x) = -4\left(x^2 - 9\right)^{-2}(2x) = \dfrac{-8x}{\left(x^2 - 9\right)^2}$ Find the first derivative

$$f''(x) = \frac{(x^2-9)^2(-8) - (-8x)(2)(x^2-9)(2x)}{(x^2-9)^4}$$ Find the second derivative using the Quotient Rule

$$= \frac{-8(x^2-9)[x^2-9-4x^2]}{(x^2-9)^4}$$ Factor $-8(x^2-9)$ out of the numerator

$$= \frac{-8(-3x^2-9)}{(x^2-9)^3}$$ Reduce the fraction

The numerator can never equal 0. Therefore the only places where the concavity might change would be where the denominator equals 0 (even though these are not values in the domain of the original function).

Test for changes in concavity around the vertical asymptotes $(x = \pm 3)$.

When $x < -3$, $f''(x) > 0$.

When $-3 < x < 3$, $f''(x) < 0$.

When $x > 3$, $f''(x) > 0$.

$f(x)$ is concave upward on $(-\infty, -3)$ and $(3, \infty)$, and concave downward on $(-3, 3)$. **(Concavity)**

32. $f'(x) = 3x^2 - 1$ Find the first derivative

$f''(x) = 6x$ Find the second derivative

$6x = 0$ Set the second derivative equal to 0

$x = 0$ Divide both sides of the equation by 6

For $x = 0$ to be the x-coordinate of an inflection point, the concavity must change on the left and right of $x = 0$.

When $x < 0$, $f''(x) < 0$.

When $x > 0$, $f''(x) > 0$.

Therefore, $(0,0)$ is an inflection point. **(Inflection Points)**

33. See #30. Since the concavity changes at $x = \frac{-\sqrt{30}}{6}$ and at $x = \frac{\sqrt{30}}{6}$, $\left(\frac{-\sqrt{30}}{6}, \frac{91}{36}\right)$ and $\left(\frac{\sqrt{30}}{6}, \frac{91}{36}\right)$ are both inflection points. **(Inflection Points)**

34. $f'(x) = 8x - 3$ Find the first derivative

$f''(x) = 8$ Find the second derivative

There is no x-value that makes $f''(x) = 0$ and therefore no inflection points. **(Inflection Points)**

35. See #31 and note that the concavity changes only around the vertical asymptotes. Therefore, there are no inflection points. **(Inflection Points)**

36. $f'(x) = 3x^2 - 8x$ Find the first derivative

$3x^2 - 8x = 0$ Set the first derivative equal to 0

$x(3x - 8) = 0$ Factor

$x = 0$ or $3x - 8 = 0$ Set each factor equal to 0 and solve

Therefore, the critical values are $x = 0, \dfrac{8}{3}$

$f''(x) = 6x - 8$ Find the second derivative

$f''(0) = -8 \Rightarrow$ relative maximum   Test $x = 0$ in the second derivative

$f''\left(\dfrac{8}{3}\right) = +8 \Rightarrow$ relative minimum   Test $x = 8/3$ in the second derivative

Therefore, $(0, 1)$ is a relative maximum. $\left(\dfrac{8}{3}, -\dfrac{229}{27}\right)$ is a relative minimum. **(Relative Extrema)**

37. $f'(x) = 4x^3 - 10x$ Find the first derivative

$2x(2x^2 - 5) = 0$ Set the first derivative equal to 0; factor the left side

$2x = 0$ or $2x^2 - 5 = 0$ Set each factor equal to 0

$x = 0$ or $x = \pm\sqrt{\dfrac{5}{2}}$ Solve each equation

$f''(x) = 12x^2 - 10$ Find the first derivative

$f''(0) < 0,$ Test $x = 0$ in the second derivative

$f''\left(\sqrt{\dfrac{5}{2}}\right) > 0,$ Test $x = \sqrt{\dfrac{5}{2}}$ in the second derivative

$f''\left(-\sqrt{\dfrac{5}{2}}\right) > 0,$ Test $x = -\sqrt{\dfrac{5}{2}}$ in the second derivative

Therefore, $(0, 4)$ is a relative maximum, $\left(\sqrt{\dfrac{5}{2}}, -\dfrac{9}{4}\right)$ is a relative minimum, and $\left(-\sqrt{\dfrac{5}{2}}, -\dfrac{9}{4}\right)$ is a relative mini-

mum. **(Relative Extrema)**

38. $f'(x) = 3x^2 - 2x - 1$ Find the first derivative

$3x^2 - 2x - 1 = 0$ Set the first derivative equal to 0

$(3x + 1)(x - 1) = 0$ Factor the left side

$x = -\dfrac{1}{3}$ or $x = 1$ Set each factor equal to 0 and solve for x

$f''(x) = 6x - 2$ Find the second derivative

$f''\left(-\dfrac{1}{3}\right) < 0$,   Test $x = -\dfrac{1}{3}$ in the second derivative

$f''(1) > 0$ Test $x = 1$ in the second derivative

Therefore, $\left(-\dfrac{1}{3}, -\dfrac{49}{27}\right)$ is a relative maximum, and $(1, -3)$ is a relative minimum. **(Relative Extrema)**

39. $+\infty$ **(Infinite Limits)**

40. $-\infty$ **(Infinite Limits)**

41. $-\infty$ **(Infinite Limits)**

42. $-\infty$ **(Infinite Limits)**

43. $+\infty$ **(Infinite Limits)**

44. a) $(-\infty, \infty)$

    b) $(0, 1)$ is the y-intercept and $(-1, 0)$ is the x-intercept

    c) No symmetry

    d) No asymptotes

    e) $f'(x) = 2x + 2$, $f''(x) = 2$. $(-1, 0)$ is a relative minimum.

    f) Increasing on $(-1, \infty)$, decreasing on $(-\infty, -1)$

    g) Concave up on $(-\infty, \infty)$

    h) No inflection points

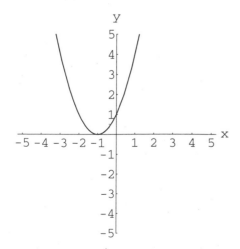

   (Graphing)

45. a) All reals, $x \neq \pm 2$

    b) $(0, 0)$

c) Symmetric with respect to the origin

d) Vertical asymptotes: $x = \pm 2$ horizontal asymptotes: $y = 0$

e) $f'(x) = \dfrac{(x^2 - 4)6 - 6x(2x)}{(x^2 - 4)^2} = \dfrac{-6x^2 - 24}{(x^2 - 4)^2}$ No critical values, no relative extrema.

f) Note that the numerator of $f'(x)$ is always negative, the denominator is always positive, so $f(x)$ is always negative and therefore $f(x)$ is decreasing on $(-\infty, -2)$, $(-2, 2), (2, \infty)$

g) $f''(x) = \dfrac{(x^2 - 4)^2(-12x) - (-6x^2 - 24) 2(x^2 - 4)(2x)}{(x^2 - 4)^4} =$

$\dfrac{-4x(x^2 - 4)[3(x^2 - 4) + (-6x^2 - 24)]}{(x^2 - 4)^4} = \dfrac{-4x[-3x^2 - 36]}{(x^2 - 4)^3}$

$f''(x)$

$$\begin{array}{ccccccc} & - & & + & & - & & + \\ \hline & & -2 & & 0 & & 2 & \end{array}$$

$f(x)$ is concave downward on $(-\infty, -2)$ and $(0, 2)$. $f(x)$ is concave upward on $(-2, 0)$ and $(2, \infty)$.

h) $(0, 0)$

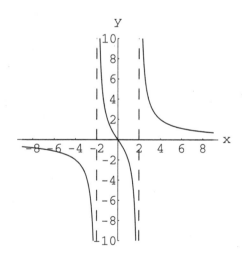

(Graphing)

46. a) $(-\infty, \infty)$

b) $(0, -2)$ $(-1, 0)$ $(2, 0)$

c) No symmetry

d) No asymptotes

e) $f'(x) = (x + 1)^2 + (x - 2) 2(x + 1)$

$= (x + 1)[(x + 1) + 2(x - 2)]$

$$= (x+1)(3x-3)$$

$$= 3(x+1)(x-1)$$

$f''(x) = 6x$ $(-1, 0)$ is a relative maximum. $(1, -4)$ is a relative minimum.

f) Increasing on $(-\infty, -1)$ and $(1, \infty)$, decreasing on $(-1, 1)$

g) $(-\infty, 0)$ concave downward, $(0, \infty)$ concave upward.

h) $(0, -2)$

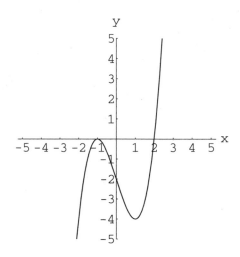

(Graphing)

47. $x + y = 48$ Write an equation to represent a "sum of two numbers is 48"

Maximize: xy Write an expression for what is to be maximized

Determine the domain: $[0, 48]$ (since the numbers must be positive)

$y = -x + 48$ Solve $x + y = 48$ for y

Maximize: $x(-x + 48) = -x^2 + 48x$ Substitute the value found for y in the expression to be maximized

$f(x) = -x^2 + 48x$ Write the function to be maximized

$f'(x) = -2x + 48$ Find the first derivative

$-2x + 48 = 0$ Set the derivative equal to 0

$x = 24$ Solve for x

Since $f''(x) = -2$, $x = 24$ gives a maximum. This is also the absolute maximum since testing x at the end-points of the interval gives a product of 0. When $x = 24$, $y = 24$. Therefore, the product is a maximum when both numbers are 24. **(Max-min Applications)**

48. Given $xy = 48$ Write an equation to represent "product is 48"

Minimize: $x + y$ Write an expression for what is to be minimized

$y = \dfrac{48}{x}$ Solve $xy = 48$ for y

Minimize $x + \dfrac{48}{x}$ Substitute the value found for y in the expression to be minimized

$f(x) = x + 48x^{-1}$ Write the function to be minimized

$f'(x) = 1 - 48x^{-2} = 1 - \dfrac{48}{x^2} = \dfrac{x^2 - 48}{x^2}$ Find the first derivative

Critical values are $\pm\sqrt{48} = \pm 4\sqrt{3}$. $f''(x) = 96x^{-3}$. When $x = 4\sqrt{3}$, $f(x)$ is a minimum. When each number equals $4\sqrt{3}$, the sum is minimized. Their sum is $4\sqrt{3} + 4\sqrt{3} = 8\sqrt{3}$. (**Max-min Applications**)

49.

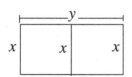

$xy = 600\text{ft}^2$ Write an equation to represent "the area is 600 square feet"

Minimize: $3x + 2y$ Write an expression for the function to be minimized

$y = \dfrac{600}{x}$ Solve $xy = 600$ for y

$f(x) = 3x + 2\left(\dfrac{600}{x}\right)$ Substitute $y = \dfrac{600}{x}$ in the expression to be minimized

$f'(x) = 3 - 1200x^{-2} = 3 - \dfrac{1200}{x^2} = \dfrac{3x^2 - 1200}{x^2}$ Find the first derivative

$3x^2 - 1200 = 0$ Set the numerator equal to 0 to find critical values

$x^2 = \dfrac{1200}{3}$ Add 1200 to both sides of the equation; divide both sides of the equation by 3

$x = \sqrt{\dfrac{1200}{3}} = \sqrt{400} = 20$ Take the square root of both sides of the equation. Use only the positive square root since a length of fence cannot be negative.

We will use the second derivative test to determine where maximums and minimums occur:

$f''(x) = 2400x^{-3}$ Find the second derivative

$f''(20) > 0 \Rightarrow x = 20$ gives a minimum. When $x = 20$ feet, $y = 30$ feet. Therefore, the minimum amount of fencing is 120 feet. (**Max-min Applications**)

Grade Yourself

Circle the question numbers that you had incorrect. Then indicate the number of questions you missed. If you answered more than three questions incorrectly, you will then have to focus on that topic. If a topic has less than three questions and you had at least one wrong, we suggest you study that topic also. Read your textbook, a review book, or ask your teacher for help.

Subject: Applications of Derivatives

Topic	Question Numbers	Number Incorrect
Related Rates	1, 2, 3, 4, 5, 6, 7, 8, 9, 10	
Differentials	11, 12, 13, 14, 15	
Increasing and Decreasing Functions	16, 17, 18, 19, 20, 21	
Absolute Extrema	22, 23, 24	
Relative Extrema	25, 26, 27, 36, 37, 38	
Concavity	28, 29, 30, 31	
Inflection Points	32, 33, 34, 35	
Infinite Limits	39, 40, 41, 42, 43	
Sophisticated Graphing	44, 45, 46	
Max-Min Applications	47, 48, 49	

Integrals

Brief Yourself

Integration is the reverse process of differentiation. In integration, you are given the derivative of a function, and are asked to find the original function. Although you may be allowed access to Tables of Integrals, or to symbolic integration with your calculator, memorizing the following list of formulas will give you speed and ease of manipulation throughout the remainder of your study of calculus:

$$\int k\,dx = kx + C\,, k, C \text{ constants}$$

$$\int kf(x)\,dx = k\int f(x)\,dx$$

$$\int [f(x) \pm g(x)]\,dx = \int f(x)\,dx \pm \int g(x)\,dx$$

$$\int u^n du = \frac{u^{n+1}}{n+1} + C\,, n \neq -1 \quad \text{(Power Rule)}$$

$$\int \cos u\,du = \sin u + C$$

$$\int \sin u\,du = -\cos u + C$$

$$\int \sec^2 u\,du = \tan u + C$$

$$\int_a^b f(x)\,dx = F(x)\Big|_a^b = F(b) - F(a) \text{ where } F'(x) = f(x) \quad \text{(The Fundamental Theorem of Calculus)}$$

The Fundamental Theorem is used to evaluate a definite integral. *After* the antiderivative is found, substitute in the limits of integration and subtract.

Generally a chapter that introduces integration contains a theoretical development of integration as the area bounded by a curve and the *x*-axis, which can be computed by a summation process. Questions about this development are included in the test items, but your instructor may consider this material optional.

Test Yourself

Find each integral.

1. $\int 8\,dx$

2. $\int x^3\,dx$

3. $\int 5x\,dx$

4. $\int (x^2 + x^5)\,dx$

5. $\int x^{-4}\,dx$

6. $\int \dfrac{1}{x^5}\,dx$

7. $\int 4\sqrt[3]{x}\,dx$

8. $\int \dfrac{(x^2 + x^5)}{x}\,dx$

9. $\int x\,(2x+1)^2\,dx$

10. $\int \sqrt{x}\,(3\sqrt{x} + x)\,dx$

11. $\int 6\sin x\,dx$

12. $\int 4\cos x\,dx$

13. $\int (3\sin x - 2\cos x)\,dx$

14. $\int \dfrac{(1 + \cot x)}{\csc x}\,dx$

15. $\int \left(x^2 - 2\sec^2 x \right) dx$

16. Expand $\displaystyle\sum_{i=1}^{3} 2i$

17. Expand $\displaystyle\sum_{k=3}^{5} k^2$

18. Expand $\displaystyle\sum_{j=1}^{4} \dfrac{1}{n}(3j+1)$

19. Find the sum: $\displaystyle\sum_{i=1}^{100} 3i$

20. Find the sum: $\displaystyle\sum_{i=1}^{50} (i^2 + 2)$

21. Find the sum: $\displaystyle\sum_{i=1}^{100} 4$

22. Find $\displaystyle\lim_{n \to \infty} \sum_{i=1}^{n} \dfrac{3i}{n^2}$

23. Find $\displaystyle\lim_{n \to \infty} \sum_{i=1}^{n} \dfrac{2}{n}$

24. Find $\displaystyle\lim_{n \to \infty} \sum_{i=1}^{n} \dfrac{i^2}{n^3}$

25. Use 4 circumscribed rectangles of equal width to approximate the area bounded by $y = 2x + 3$ and the x-axis on $[0, 2]$.

26. Use 8 inscribed rectangles to approximate the area bounded by $y = \sqrt{x}$ and the x-axis on $[0, 2]$.

27. If $f(x) = x^2 + 1$ and we wish to approximate the area bounded by the graph of f and the x-axis on $[0, 4]$ using upper sums (or circumscribed rectangles) with 8 subintervals, find:

   a)   the width of each subinterval

   b)   the height of the first rectangle

   c)   the area of the first rectangle

28. Use the limit process to find the area of the region bounded by $y = 3x + 1$ and the x-axis over the interval $[0, 1]$.

29. Use the limit process to find the area of the region bounded by $y = x^2 + 2$ and the x-axis over the interval $[0, 1]$.

30. Use the limit process to find the area of the region bounded by $y = -x^2 + 1$ and the x-axis over the interval $[-1, 1]$.

Evaluate each definite integral using the Fundamental Theorem of Calculus.

31. $\int_0^1 (x^2 + 2)\, dx$

32. $\int_0^1 (3x + 1)\, dx$

33. $\int_{-1}^1 (-x^2 + 1)\, dx$

34. $\int_4^{16} 3\sqrt{x}\, dx$

35. $\int_2^2 (x^3 + 6x^2 + 2x - 3)\, dx$

36. $\int_0^{\pi/2} \cos x\, dx$

37. $\int_0^{\pi} \sin x\, dx$

38. Find the area bounded by $f(x) = 2x - x^2$, the x-axis, and the lines $x = 1$ and $x = 1.5$.

39. Find the area bounded by $f(x) = \sqrt{x} + 2$, the x-axis, and the lines $x = 1$ and $x = 4$.

40. Find the area bounded by $f(x) = x^2 - 4$, the x-axis, and the lines $x = -1$ and $x = 1$.

41. Find the area bounded by $f(x) = -2\cos x$, the x-axis, and the lines $x = 0$ and $x = \pi/4$.

42. Find the area bounded by $f(x) = x^2 + 4x + 3$, the x-axis, and the lines $x = -2$ and $x = 0$.

43. Find the area bounded by $y = 4\sin x$, the x-axis, and the lines $x = -\pi/4$ and $x = \pi$.

44. Evaluate $\int (x^2 + 3x)^6 (2x + 3)\, dx$

45. Evaluate $\int 4\sin 4x\, dx$

46. Evaluate $\int (x^3 + 6)^7 x^2\, dx$

47. Evaluate $\int \sqrt{3x - 2}\, dx$

48. Evaluate $\int \cos 2x\, dx$

49. Evaluate $\int \dfrac{x}{(x^2 + 5)^3}\, dx$

50. Evaluate $\int \sin^3 x \cos x\, dx$

51. Evaluate $\int \cos^2 x \sin x\, dx$

52. Evaluate $\int \sin^4 2x \cos 2x\, dx$

53. Evaluate $\int_1^2 \dfrac{x + 1}{(x^2 + 2x)^3}\, dx$

54. Evaluate $\int_0^{\pi/2} \sin^2 x \cos x\, dx$

Check Yourself

1. $8x + C$ (**Antiderivatives**)

2. $\dfrac{x^4}{4} + C$ (**Antiderivatives**)

3. $\displaystyle\int 5x\,dx = 5\int x\,dx$ Factor out the constant

       $= \dfrac{5x^2}{2} + C$ Use the Power Rule

       $= \dfrac{5}{2}x^2 + C$ The answer may be written in either form   (**Antiderivatives**)

4. $\displaystyle\int (x^2 + x^5)\,dx = \int x^2\,dx + \int x^5\,dx$ The integral of a sum is the sum of the integrals

       $= \dfrac{x^3}{3} + \dfrac{x^6}{6} + C$ Use the Power Rule on each term   (**Antiderivatives**)

5. $\dfrac{x^{-3}}{-3} + C$ or $-\dfrac{1}{3}x^{-3} + C$ or $-\dfrac{1}{3x^3} + C$ Be careful with the exponent. $-4 + 1 = -3$ (**Antiderivatives**)

6. $\displaystyle\int \dfrac{1}{x^5}\,dx = \int x^{-5}\,dx$ Write with a negative exponent so the Power Rule can be used

       $= \dfrac{x^{-4}}{-4} + C$ Use $\displaystyle\int u^n\,du = \dfrac{u^{n+1}}{n+1} + C$

       $= -\dfrac{1}{4x^4} + C$ Answers are usually written without negative exponents   (**Antiderivatives**)

7. $\displaystyle\int 4\sqrt[3]{x}\,dx = 4\int x^{1/3}\,dx$ Factor out the constant, 4; write the radical as a fractional exponent

       $= 4\left[\dfrac{x^{4/3}}{4/3}\right] + C$ Use $\displaystyle\int u^n\,du = \dfrac{u^{n+1}}{n+1} + C$

       $= 3x^{4/3} + C$ $4 \div \dfrac{4}{3} = 4 \cdot \dfrac{3}{4} = 3$ (**Antiderivatives**)

8. There is no quotient rule for integration. Simplify first:

   $\displaystyle\int \dfrac{x^2 + x^5}{x}\,dx = \int (x + x^4)\,dx$ Divide x into x^2 and x^5

       $= \dfrac{x^2}{2} + \dfrac{x^5}{5} + C$ Use $\displaystyle\int [f(x) + g(x)]\,dx = \int f(x)\,dx + \int g(x)\,dx$ (**Antiderivatives**)

9. Simplify first by multiplying:

$$\int x\,(2x+1)^2 dx \;=\; \int x\,(4x^2+4x+1)\,dx \qquad \text{Multiply } (2x+1)^2 \text{ first}$$

$$=\; \int (4x^3+4x^2+x)\,dx \qquad \text{Distribute the } x \text{ to each term in the parentheses}$$

$$=\; x^4 + \frac{4x^3}{3} + \frac{x^2}{2} + C \qquad \text{Use } \int [f(x)+g(x)]\,dx \;=\; \int f(x)\,dx + \int g(x)\,dx$$

(Antiderivatives)

10. $\displaystyle\int \sqrt{x}\,(3\sqrt{x}+x)\,dx \;=\; \int (3x + x\sqrt{x})\,dx \qquad \text{Multiply each term by } \sqrt{x}$

$$=\; \int (3x + x^{3/2})\,dx \qquad \text{Write } x\sqrt{x} \;=\; x \cdot x^{1/2} \;=\; x^{3/2}$$

$$=\; \frac{3x^2}{2} + \frac{2x^{5/2}}{5} + C \qquad \text{Use } \int [f(x)+g(x)]\,dx \;=\; \int f(x)\,dx + \int g(x)\,dx$$

(Antiderivatives)

11. $-6\cos x + C$ **(Antiderivatives)**

12. $4\sin x + C$ **(Antiderivatives)**

13. $-3\cos x - 2\sin x + C$ **(Antiderivatives)**

14. $\displaystyle\int \frac{1+\cot x}{\csc x}\,dx \;=\; \int \left(\frac{1}{\csc x} + \frac{\cot x}{\csc x} \right) dx \qquad \text{Divide each term in the numerator by } \csc x$

$$=\; \int \frac{1}{\csc x}\,dx + \int \frac{\cot x}{\csc x}\,dx \qquad \text{Use } \int [f(x)+g(x)]\,dx \;=\; \int f(x)\,dx + \int g(x)\,dx$$

$$=\; \int \sin x\,dx + \int \left(\frac{\cos x}{\sin x} \cdot \sin x \right) dx \qquad \text{Use the identities } \cot x \;=\; \frac{\cos x}{\sin x} \text{ and } \frac{1}{\csc x} \;=\; \sin x$$

$$=\; \int \sin x\,dx + \int \cos x\,dx \;=\; -\cos x + \sin x + C \quad \textbf{(Antiderivatives)}$$

15. $\dfrac{x^3}{3} - 2\tan x + C$ **(Antiderivatives)**

16. $2(1) + 2(2) + 2(3) = 12$ **(Summation Notation)**

17. $3^2 + 4^2 + 5^2 = 50$ **(Summation Notation)**

18. $\displaystyle\sum_{j=1}^{4} \frac{1}{n}(3j+1) \;=\; \frac{1}{n}\sum_{j=1}^{4}(3j+1) \qquad \frac{1}{n}$ can be factored out as a constant (it does not contain the index j)

$$= \frac{1}{n} \left(3\,(1) + 1 + 3\,(2) + 1 + 3\,(3) + 1 + 3\,(4) + 1 \right) \qquad \text{Expand using the definition of } \Sigma$$

$$= \frac{1}{n}\,(34) \; = \; \frac{34}{n} \quad \textbf{(Summation Notation)}$$

19. $\displaystyle\sum_{i=1}^{100} 3i = 3 \sum_{i=1}^{100} i \quad$ Factor out the constant, 3

$$= 3\frac{(100)\,(101)}{2} \quad \text{Use } \sum_{i=1}^{n} i = \frac{n\,(n+1)}{2}$$

$$= 15{,}150 \qquad \text{Simplify the fraction} \quad \textbf{(Summation Notation)}$$

20. $\displaystyle\sum_{i=1}^{50} (i^2 + 2) = \sum_{i=1}^{50} i^2 + \sum_{i=1}^{50} 2 \quad$ Use $\displaystyle\sum (a+b) = \sum a + \sum b$

$$= \frac{50\,(51)\,(101)}{6} + 2\,(50) \qquad \text{Use } \sum_{i=1}^{n} i^2 = \frac{n\,(n+1)\,(2n+1)}{6} \text{ and } \sum_{i=1}^{n} c = cn\,, c \text{ a constant}$$

$$= 43{,}025 \quad \textbf{(Summation Notation)}$$

21. $\displaystyle\sum_{i=1}^{100} 4 = 400 \qquad \text{Use } \sum_{i=1}^{n} c = cn\,, c \text{ a constant} \quad \textbf{(Summation Notation)}$

22. $\displaystyle\lim_{n\to\infty} \sum_{i=1}^{n} \frac{3i}{n^2} = \lim_{n\to\infty} \frac{3}{n^2} \sum_{i=1}^{n} i \quad$ Factor out the constant $\dfrac{3}{n^2}$

$$= \lim_{n\to\infty} \left[\frac{3}{n^2} \frac{n\,(n+1)}{2} \right] \qquad \text{Use } \sum_{i=1}^{n} i = \frac{n\,(n+1)}{2}$$

$$= \lim_{n\to\infty} \left[\frac{3n^2 + 3n}{2n^2} \right] = \lim_{n\to\infty} \left[\frac{3n^2}{2n^2} + \frac{3n}{2n^2} \right] \qquad \text{Simplify inside the brackets}$$

$$= \lim_{n\to\infty} \left(\frac{3}{2} + \frac{3}{2n} \right) = \frac{3}{2} \text{ since } \frac{3}{2n} \to 0 \text{ as } n\to\infty \quad \textbf{(Infinite Limits of Sums)}$$

23. $\displaystyle\lim_{n\to\infty} \frac{2}{n} \sum_{i=1}^{n} 1 = \lim_{n\to\infty} \left(\frac{2}{n} \cdot n \right) = 2 \quad \textbf{(Infinite Limits of Sums)}$

24. $\displaystyle\lim_{n\to\infty} \frac{1}{n^3} \sum_{i=1}^{n} i^2 = \lim_{n\to\infty} \frac{1}{n^3} \cdot \frac{n\,(n+1)\,(2n+1)}{6} = \lim_{n\to\infty} \left[\frac{1}{n^3} \cdot \frac{2n^3 + 3n^2 + n}{6} \right] = \lim_{n\to\infty} \left(\frac{2}{6} + \frac{3}{6n} + \frac{1}{6n^2} \right) = \frac{1}{3}$

$\textbf{(Infinite Limits of Sums)}$

25.

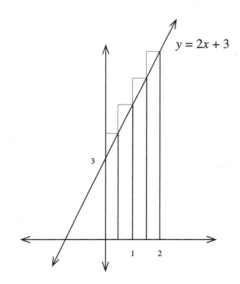

$\Delta x = \dfrac{2-0}{4} = \dfrac{2}{4} = \dfrac{1}{2}$. Thus, the intervals are $\left[0, \dfrac{1}{2}\right], \left[\dfrac{1}{2}, 1\right], \left[1, \dfrac{3}{2}\right], \left[\dfrac{3}{2}, 2\right]$. Using $f(x) = 2x + 3$ and the right endpoint of each subinterval to find the height of each rectangle, we have:

$$\text{Area} \approx \dfrac{1}{2}f\left(\dfrac{1}{2}\right) + \dfrac{1}{2}f(1) + \dfrac{1}{2}f\left(\dfrac{3}{2}\right) + \dfrac{1}{2}f(2)$$

$$= \dfrac{1}{2}\left[2\left(\dfrac{1}{2}\right) + 3 + 2(1) + 3 + 2\left(\dfrac{3}{2}\right) + 3 + 2(2) + 3\right]$$

$$= 11 \quad \textbf{(Approximating Area)}$$

26.

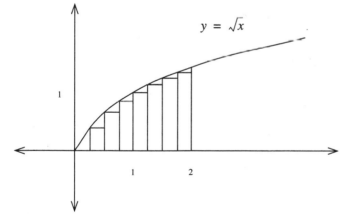

$\Delta x = \dfrac{2-0}{8} = \dfrac{1}{4} = 0.25$. Thus the intervals are $[0, .25], [.25, .5], [.5, .75], [.75, 1], [1, 1.25], [1.25, 1.5],$ $[1.5, 1.75], [1.75, 2]$. Using the left endpoint of each subinterval to find the height of each rectangle, we have:

$$\text{Area} \approx 0.25\,[\sqrt{0} + \sqrt{0.25} + \sqrt{0.5} + \sqrt{0.75} + \sqrt{1} + \sqrt{1.25} + \sqrt{1.5} + \sqrt{1.75}] = 1.68 \quad \textbf{(Approximating Area)}$$

27. a) $\Delta x = \dfrac{4-0}{8} = \dfrac{1}{2}$

b) $f\left(\dfrac{1}{2}\right) = \left(\dfrac{1}{2}\right)^2 + 1 = \dfrac{5}{4}$

c) Area = width × height = $\frac{1}{2}\left(\frac{5}{4}\right) = \frac{5}{8}$

(Approximating Area)

28. Area = $\lim\limits_{n \to \infty} \sum\limits_{i=1}^{n} f(C_i)\,\Delta x$, where $\Delta x = \dfrac{b-a}{n} = \dfrac{1-0}{n} = \dfrac{1}{n}$. Let C_i be the right endpoint of each

subinterval so that $C_i = \dfrac{i}{n}$. Then

Area = $\lim\limits_{n \to \infty} \sum\limits_{i=1}^{n} \left[3\left(\dfrac{i}{n}\right)+1\right]\dfrac{1}{n}$

$= \lim\limits_{n \to \infty} \sum\limits_{i=1}^{n} \left[\dfrac{3i}{n^2}+\dfrac{1}{n}\right]$ Multiply by $\dfrac{1}{n}$

$= \lim\limits_{n \to \infty} \left(\dfrac{3}{n^2}\sum\limits_{i=1}^{n} i + \dfrac{1}{n}\sum\limits_{i=1}^{n} 1\right)$ Factor out the constants, $\dfrac{3}{n^2}$ and $\dfrac{1}{n}$

$= \lim\limits_{n \to \infty} \left[\dfrac{3}{n^2}\left(\dfrac{n(n+1)}{2}\right)+\dfrac{1}{n}(n)\right]$ Use $\sum\limits_{i=1}^{n} i = \dfrac{n(n+1)}{2}$ and $\sum\limits_{i=1}^{n} c = cn$

$= \lim\limits_{n \to \infty} \left[\dfrac{3}{2}\left(\dfrac{n^2+n}{n^2}\right)+1\right]$ Simplify the fractions

$= \lim\limits_{n \to \infty} \left[\dfrac{3}{2}\left(1+\dfrac{1}{n}\right)+1\right] = \dfrac{3}{2}+1 = \dfrac{5}{2}$ Take the limit as $n \to \infty$

(Finding Area using the Limit Definition)

29. $\Delta x = \dfrac{1-0}{n} = \dfrac{1}{n}$. Let C_i be the right end point of each subinterval so that $C_i = \dfrac{i}{n}$.

Area = $\lim\limits_{n \to \infty} \sum\limits_{i=1}^{n} \left[\left(\dfrac{i}{n}\right)^2+2\right]\dfrac{1}{n}$ Use the formula for area

$= \lim\limits_{n \to \infty} \sum\limits_{i=1}^{n} \left[\dfrac{i^2}{n^3}+\dfrac{2}{n}\right]$ Multiply by $\dfrac{1}{n}$

$= \lim\limits_{n \to \infty} \left(\dfrac{1}{n^3}\sum\limits_{i=1}^{n} i^2 + \dfrac{2}{n}\sum\limits_{i=1}^{n} 1\right)$ Factor out the constants $\dfrac{1}{n^3}$ and $\dfrac{2}{n}$

$= \lim\limits_{n \to \infty} \left[\dfrac{1}{n^3}\left(\dfrac{n(n+1)(2n+1)}{6}\right)+\dfrac{2}{n}(n)\right]$ Use $\sum\limits_{i=1}^{n} i = \dfrac{n(n+1)}{2}$ and $\sum\limits_{i=1}^{n} c = cn$

$$= \lim_{n \to \infty} \left[\frac{1}{n^3} \left(\frac{2n^3 + 3n^2 + n}{6} \right) + 2 \right] \qquad \text{Simplify inside the brackets}$$

$$= \lim_{n \to \infty} \left[\frac{2}{6} + \frac{3}{6n} + \frac{1}{6n^2} + 2 \right] = \frac{2}{6} + 2 = \frac{7}{3} \qquad \text{Take the limit as } n \to \infty$$

(Finding Area using the Limit Definition)

30. $\Delta x = \dfrac{1 - (-1)}{n} = \dfrac{2}{n}$. Let C_i be the right end point of each subinterval so that $C_i = -1 + \dfrac{2i}{n}$. Then

$$\text{Area} = \lim_{n \to \infty} \sum_{i=1}^{n} \left[- \left(-1 + \frac{2i}{n} \right)^2 + 1 \right] \frac{2}{n} \qquad \text{Use the formula for area}$$

$$= \lim_{n \to \infty} \sum_{i=1}^{n} \left[- \left(1 - \frac{4i}{n} + \frac{4i^2}{n^2} \right) + 1 \right] \frac{2}{n} \qquad \text{Multiply } \left(-1 + \frac{2i}{n} \right)^2$$

$$= \lim_{n \to \infty} \sum_{i=1}^{n} \left(\frac{8i}{n^2} - \frac{8i^2}{n^3} \right) \qquad \text{Combine similar terms; multiply by } \frac{2}{n}$$

$$= \lim_{n \to \infty} \left[\frac{8}{n^2} \sum i - \frac{8}{n^3} \sum i^2 \right] \qquad \text{Factor out the constants } \frac{8}{n^2} \text{ and } \frac{8}{n^3}$$

$$= \lim_{n \to \infty} \left[\frac{8}{n^2} \frac{n(n+1)}{2} - \frac{8}{n^3} \frac{n(n+1)(2n+1)}{6} \right] \qquad \text{Use } \sum_{i=1}^{n} i^2 = \frac{n(n+1)(2n+1)}{6} \text{ and}$$

$$\sum_{i=1}^{n} i = \frac{n(n+1)}{2}$$

$$= \lim_{n \to \infty} \left[4 + \frac{4}{n} - \frac{16}{6} - \frac{24}{6n} - \frac{8}{6n^2} \right] = 4 - \frac{16}{6} = \frac{4}{3} \qquad \text{Take the limit as } n \to \infty$$

(Finding Area using the Limit Definition)

31. $\left(\dfrac{x^3}{3} + 2x \right) \Big|_0^1 = \dfrac{1}{3} + 2(1) - 0 = \dfrac{7}{3}$ (See problem #29) **(Fundamental Theorem of Calculus)**

32. $\left(\dfrac{3x^2}{2} + x \right) \Big|_0^1 = \dfrac{3}{2} + 1 - 0 = \dfrac{5}{2}$ (See problem #28) **(Fundamental Theorem of Calculus)**

33. $\left(-\dfrac{x^3}{3} + x \right) \Big|_{-1}^1 = \left(-\dfrac{1}{3} + 1 \right) - \left(\dfrac{1}{3} - 1 \right) = \dfrac{2}{3} - \left(-\dfrac{2}{3} \right) = \dfrac{4}{3}$ (See problem #30) **(Fundamental Theorem of Calculus)**

34. $\int_4^{16} 3\sqrt{x}\,dx = 3\int_4^{16} x^{1/2}\,dx = 3\left[\frac{2}{3}x^{3/2}\Big|_4^{16}\right] = 2[16^{3/2} - 4^{3/2}] = 112$ **(Fundamental Theorem of Calculus)**

35. $\int_2^2 (x^3 + 6x^2 + 2x - 3)\,dx = 0$. Note that the limits of integration are equal and therefore you do not need to actually perform the integration to get the answer 0. **(Properties of Definite Integrals)**

36. $\sin x\big|_0^{\pi/2} = \sin\left(\frac{\pi}{2}\right) - \sin 0 = 1 - 0 = 1$ **(Fundamental Theorem of Calculus)**

37. $-\cos x\big|_0^{\pi} = -\cos\pi - (-\cos 0) = -(-1) - (-1) = 2$ **(Fundamental Theorem of Calculus)**

38. The graph of $f(x) = 2x - x^2$ is above the x-axis on $[1, 1.5]$, so area =

$\int_1^{1.5} (2x - x^2)\,dx = \left(x^2 - \frac{x^3}{3}\right)\Big|_1^{1.5} \approx (1.125 - 0.667) \approx 0.458$ **(Area using Definite Integrals)**

39. The graph of $f(x) = \sqrt{x} + 2$ is above the x-axis, so area =

$\int_1^4 (\sqrt{x} + 2)\,dx = \left(\frac{2}{3}x^{3/2} + 2x\right)\Big|_1^4 = \left(\frac{2}{3}\cdot 8 + 8\right) - \left(\frac{2}{3} + 2\right) = \frac{32}{3}$ **(Area using Definite Integrals)**

40. The graph of $y = x^2 - 4$ is below the x-axis on $[-1, 1]$, so area =

$-\int_{-1}^1 (x^2 - 4)\,dx = -\left(\frac{x^3}{3} - 4x\right)\Big|_{-1}^1 = -\left(\frac{1}{3} - 4\right) - \left(-\left(-\frac{1}{3} + 4\right)\right) = \frac{22}{3}$ **(Area using Definite Integrals)**

41. Sketch the graph of $f(x) = -2\cos x$ to see that it is below the x-axis on $\left[0, \frac{\pi}{4}\right]$. Area =

$-\int_0^{\pi/4} (-2\cos x)\,dx = 2\sin x\big|_0^{\pi/4} = 2\sin\frac{\pi}{4} - 2\sin 0 = \sqrt{2}$ **(Area using Definite Integrals)**

42. Sketch the graph of $y = x^2 + 4x + 3$. The graph lies below the x-axis on $[-2, -1]$ and above the x-axis on

$[-1, 0]$. The area is $-\int_{-2}^{-1} (x^2 + 4x + 3)\,dx + \int_{-1}^0 (x^2 + 4x + 3)\,dx = -\left(\frac{x^3}{3} + 2x^2 + 3x\right)\Big|_{-2}^{-1} +$

$\left(\frac{x^3}{3} + 2x^2 + 3x\right)\Big|_{-1}^0 = 2$ **(Area using Definite Integrals)**

43. The graph of $y = 4\sin x$ lies below the x-axis on $\left[-\frac{\pi}{4}, 0\right]$ and above the x-axis on $[0, \pi]$.

$-\int_{-\pi/4}^0 4\sin x\,dx + \int_0^{\pi} 4\sin x\,dx = 4\cos x\big|_{-\pi/4}^0 - 4\cos x\big|_0^{\pi} = 12 - 2\sqrt{2}$ **(Area using Definite Integrals)**

44. Let $u = x^2 + 3x$

then $du = (2x + 3)\,dx$

The integral can then be written as:

$$\int u^6 du = \frac{u^7}{7} + C = \frac{(x^2 + 3x)^7}{7} + C$$

(Integration using Substitution)

45. Let $u = 4x$, then $du = 4dx$.

The integral can then be written $\int \sin u\, du = -\cos u + C = -\cos 4x + C$

(Integration using Substitution)

46. $\int (x^3 + 6)^7 x^2 dx = \frac{1}{3}\int (x^3 + 6)^7 3x^2 dx = \frac{1}{3} \cdot \frac{(x^3 + 6)^8}{8} + C = \frac{1}{24}(x^3 + 6)^8 + C$ **(Integration using Substitution)**

47. $\int (3x - 2)^{1/2} dx = \frac{1}{3}\int (3x - 2)^{1/2} 3dx = \frac{1}{3} \cdot \frac{2}{3}(3x - 2)^{3/2} + C = \frac{2}{9}(3x - 2)^{3/2} + C$ **(Integration using Substitution)**

48. $\int \cos 2x\, dx = \frac{1}{2}\int \cos 2x\,(2dx) = \frac{1}{2}\sin 2x + C$ **(Integration using Substitution)**

49. $\int (x^2 + 5)^{-3} x\, dx = \frac{1}{2}\int (x^2 + 5)^{-3} 2x\, dx = \frac{1}{2} \cdot \frac{(x^2 + 5)^{-2}}{-2} + C = -\frac{1}{4(x^2 + 5)^2} + C$ **(Integration using Substitution)**

50. $\int \sin^3 x \cos x\, dx = \int (\sin x)^3 \cos x\, dx$.

Let $u = \sin x$ then $du = \cos x\, dx$.

$$\int (\sin x)^3 \cos x\, dx = \int u^3 du = \frac{u^4}{4} + C \quad \text{Use } u\text{-substitution}$$

$$= \frac{\sin^4 x}{4} + C \quad \text{Replace } u \text{ with } \sin x$$

(Integration using Substitution)

51. $\int \cos^2 x \sin x\, dx = \int (\cos x)^2 \sin x\, dx$

Let $u = \cos x$. Then $du = -\sin x\, dx$.

$$\int (\cos x)^2 \sin x\, dx = -\int u^2 du = -\frac{u^3}{3} + C \quad \text{Use } u\text{-substitution}$$

$$= -\frac{\cos^3 x}{3} + C \quad \text{Replace } u \text{ with } \cos x$$

(Integration using Substitution)

52. Let $u = \sin 2x$. Then $du = 2\cos 2x dx$.

$$\frac{1}{2}\int \sin^4 2x \,(2\cos 2x)\, dx = \frac{1}{2}\int u^4 du = \frac{1}{2}\cdot\frac{u^5}{5} + C = \frac{1}{10}\sin^5 2x + C.$$ **(Integration using Substitution)**

53. Let $u = x^2 + 2x$. Then $du = (2x + 2)dx$.

$$\int_1^2 (x^2 + 2x)^{-3}(x+1)\,dx$$

$$= \frac{1}{2}\int_1^2 (x^2 + 2x)^{-3}\,2\,(x+1)\,dx$$

$$= \frac{1}{2}\int_{-3}^8 u^{-3} du.$$

We change the limits of integration:

when $x = 2$, $u = 2^2 + 2\,(2) = 8$. When $x = 1$, $u = 1^2 + 2\,(1) = 3$.

Now integrate:

$$\frac{1}{2}\int_3^8 u^{-3} du = \left(\frac{1}{2}\right)\frac{u^{-2}}{-2}\Big|_3^8 = -\frac{1}{4}\left[\frac{1}{64} - \frac{1}{9}\right] = \frac{55}{2304}$$ **(Substitution – Changing the Limits of Integration)**

54. Let $u = \sin x$, $du = \cos x dx$. When $x = \dfrac{\pi}{2}$, $u = \sin\dfrac{\pi}{2} = 1$. When $x = 0$, $u = \sin 0 = 0$. $\displaystyle\int_0^1 u^2 du = \frac{u^3}{3}\Big|_0^1 = \frac{1}{3}$

(Substitution – Changing the Limits of Integration)

Grade Yourself

Circle the question numbers that you had incorrect. Then indicate the number of questions you missed. If you answered more than three questions incorrectly, you will then have to focus on that topic. If a topic has less than three questions and you had at least one wrong, we suggest you study that topic also. Read your textbook, a review book, or ask your teacher for help.

Subject: Integrals

Topic	Question Numbers	Number Incorrect
Antiderivatives	1, 2, 3, 4, 5, 6, 7, 8, 9, 10, 11, 12, 13, 14, 15	
Summation Notation	16, 17, 18, 19, 20, 21	
Infinite Limits of Sums	22, 23, 24	
Approximating Area	25, 26, 27	
Finding Area using the Limit Definition	28, 29, 30	
Fundamental Theorem of Calculus	31, 32, 33, 34, 36, 37	
Properties of Definite Integrals	35	
Area using Definite Integrals	38, 39, 40, 41, 42, 43	
Integration using Substitution	44, 45, 46, 47, 48, 49, 50, 51, 52	
Substitution—Changing the Limits of Integration	53, 54	

Applications of Integration

6

Brief Yourself

The questions in this chapter involve using integration to solve problems. The applications tested include finding the area between two curves, volumes of solids of revolution, arc length, area of a surface of revolution, work, fluid force, and finding moments and center of mass.

Whenever possible, use a sketch to help you set up the problem. Also, use the principle of examining a slice and then summing the results to set up the necessary integrals.

When finding the area bounded by two curves, sketch the graphs on one set of axes (or use your graphing calculator to draw both curves). Find the intersection points of the curves by setting each equation equal to y (or x) and then setting them equal to each other. If the curves intersect in more than two points, be careful to set up separate integrals for each region between the points of the intersection. Be sure to note which curve is the top curve and which curve is the bottom curve between the points of intersection when using dx slices (or which curve is the right curve and which curve is the left curve when using dy slices).

If you use dx slices to find the area bounded by two curves, use x values as limits of integration. When finding the area bounded by two curves using dx slices, subtract the function for the bottom curve *from* the function for the top curve (Top Curve – Bottom Curve) to find the height of a rectangular slice. Note that to use dx slices, each equation should be solved for y. If you use dy slices to find the area bounded by two curves, use y values as limits of integration. When finding the area bounded by two curves using dy slices, subtract the function for the left curve *from* the function for the right curve (Right Curve – Left Curve) to find the height of a rectangular slice. Note that to use dy slices, each equation should be solved for x.

Some formulas needed for this chapter include:

Volume of solid of revolution using the Washer Method: $V = \pi \int_a^b ([R(x)]^2 - [r(x)]^2) \, dx$ where $R(x)$

represents the region bounded by the outer radius and $r(x)$ represents the region bounded by the inner radius. Note that the disc method is a variation of this formula where the inner radius is 0 and the forumla

becomes $V = \pi \int_a^b [R(x)]^2 dx$

Volume of solid of revolution using the Shell Method: $V = 2\pi \int_a^b r(x) h(x) \, dx$ where $r(x)$ is the radius of a

shell and $h(x)$ is the height of a shell

Arc Length: $s = \int_a^b \sqrt{1 + [f'(x)]^2}\, dx$ where the curve is given as $y = f(x)$ on $[a, b]$.

Area of a Surface of Revolution: $S = 2\pi \int_a^b r\sqrt{1 + [f'(x)]^2}\, dx$ where r represents the distance from the axis of revolution to the curve.

Work: $W = \int_a^b F(x)\, dx$ where F is the force

Fluid Force: $F = \delta \int_{y_1}^{y_2} hL\, dy$ where h is the depth of a horizontal slice L units in length

Center of Mass: $(\bar{x}, \bar{y}) = \left(\dfrac{\text{Moment about the y axis}}{\text{total mass}}, \dfrac{\text{Moment about the x axis}}{\text{total mass}} \right)$

Test Yourself

1. Find the area bounded by $f(x) = x^2$ and $g(x) = -3x - 2$.

2. Find the area bounded by $f(x) = x^2 - 2x - 1$ and $g(x) = 2x + 4$.

3. Find the area bounded by $f(x) = x^2 + 4x$ and $g(x) = 0$.

4. Find the area bounded by $f(x) = -x^2 + 4$ and $g(x) = x^2 - 4$.

5. Find the area bounded by $x = 2 - y^2$ and $y = x$.

6. Find the area bounded by $x = y^2 + 2y - 3$ and $x = 5$.

7. Find the area bounded by $f(x) = x^3 - x$ and $g(x) = 0$.

8. Find the area bounded by $f(x) = 2\sin x$ and $g(x) = 2\cos x$ on $\left[0, \dfrac{5\pi}{4} \right]$.

9. Find the area bounded by $f(x) = \sin 2x$ and $g(x) = -\sin 2x$ on $[0, \pi]$.

10. Find the volume of the solid formed by revolving the region in the first quadrant bounded by $y = 9 - x^2, x = 0, y = 0$ about the x-axis.

11. Find the volume of the solid formed by revolving the region in the first quadrant bounded by $y = 9 - x^2, x = 0, y = 0$ about the y-axis.

12. Find the volume of the solid formed by revolving the region in the first quadrant bounded by $y = 9 - x^2, x = 0, y = 3$ about the line $y = 3$.

13. Set up the integral required to find the volume of the solid formed by revolving the region in the first quadrant bounded by $y = 9 - x^2, x = 1$, $y = 0$ about the line $x = 1$.

14. Set up the integral required to find the volume of the solid formed by revolving the region in the first quadrant bounded by $y = \cos x + 1, x = 0$, $y = 0, x = \pi/2$ about the x-axis.

15. Find the volume of the solid formed by revolving the region in the first quadrant bounded by

$y = 9 - x^2, x = 0, y = 3$ about $y = 0$.

16. Find the volume of the solid formed by revolving the region in the first quadrant bounded by

    $x = 4 - y^2, x = 2, y = 0$ about the y-axis.

17. Find the volume of the solid formed by revolving the region in the first quadrant bounded by

    $y = x^2 + 2, x = 0, x = 4, y = 1$ about the x-axis.

18. Find the volume of the solid formed by revolving the region in the first quadrant bounded by

    $y = x^2, y = 4, x = 0$ about the line $y = 6$.

19. Find the volume of the solid formed by revolving the region in the first quadrant bounded by

    $y = x^2 + 4, x = 1, x = 0, y = 0$ about $x = 0$.

20. Find the volume of the solid formed by revolving the region in the first quadrant bounded by

    $y = x^2 + 4, x = 1, x = 0, y = 0$ about $x = 4$.

21. Find the volume of the solid formed by revolving the region bounded by $y = \frac{1}{2}x, y = 2, x = 0$ about the line $y = 4$.

22. Set up the integral to find the volume of the solid formed by revolving the region in the first quadrant bounded by $y = \frac{1}{x}, x = 1, y = \frac{1}{2}$ about the line $y = 2$. Use the Shell Method.

23. Set up the integral to find the volume of the solid formed by revolving the region bounded by

    $y = (x - 1)^2, x = 0, y = 0$ about $y = 0$ using

    a) the Disc Method.
    b) the Shell Method.

24. Find the arc length of $f(x) = x^{3/2} + 2$ on $[2, 4]$.

25. Find the arc length of the line segment from $(0, 0)$ to $(3, 8)$ by

    a) using the distance formula.

    b) using $L = \int_a^b \sqrt{1 + [f'(x)]^2} dx$.

26. Find the arc length of $f(x) = \frac{x^4}{8} + \frac{1}{4x^2}$ on

    $[4, 16]$.

27. Find a definite integral that represents the arc length of $y = \cos x$ on $\left[0, \frac{\pi}{4}\right]$.

28. Find a definite integral that represents the arc length of $y = \frac{1}{x}$ on $[1, 4]$.

29. Find an integral that represents the area of the surface of revolution formed by revolving

    $y = x^2$ on $[0, 1]$ about the x-axis.

30. Find the area of the surface revolution generated by revolving $y = \frac{1}{2}x^3$ on $[0, 1]$ about the x-axis.

31. Find the area of the surface revolution generated by revolving $x = \frac{1}{3}y^3$ on $0 \le y \le 1$ about the y-axis.

32. Find an integral that represents the area of the surface of revolution formed by revolving

    $y = 2\sin x$ on $\left[0, \frac{\pi}{4}\right]$ about the x-axis.

33. Find the work done in lifting a 100-pound antique desk 2 feet.

34. Find the work done in lifting a 1.4 kilogram object 0.53 meters. (Use $g = 9.8 \ m/s^2$).

35. A force of 10 pounds compresses a spring 6 inches from its natural length of 18 inches. Find the work done in compressing the spring a total of 10 inches.

36. A spring has a natural length of 40 cm. If a 30-N force is required to stretch the spring to 50 cm, find the work required to stretch the spring from 50 cm to 60 cm.

37. A cylindrical tank of diameter 20 feet and height 40 feet is filled with water. Find the work done to pump the water over the top edge of the tank.

38. A cylindrical tank of diameter 20 feet and height 40 feet is filled with water. Find the work done to pump half the water out of the tank.

39. A rectangular tank with base 5 feet by 6 feet and height 4 feet is filled with water. Find the work done to pump the water over the edge of the tank.

40. The vertical gate of a dam is the shape of a rectangle 3 feet long and 2 feet high. If the top of the gate is 1 foot below the water level, find the fluid force on the end of the gate.

41. The vertical gate of a dam is the shape of an isosceles trapezoid 10 feet across the top, 6 feet across the bottom, with a height of 4 feet. If the top of the gate is 2 feet below the water, what is the fluid force on the gate?

42. The circular window of a ship has a 2-foot diameter. If the center of the window is 10 feet below the water, set up the integral to find the fluid force on the window.

43. Find the center of mass for a system with $m_1 = 10$ at $x_1 = 3$, $m_2 = 20$ at $x_2 - 5$ and $m_3 = 30$ at $x_3 = -6$. Interpret your results.

44. Find the moment about the y-axis, the moment about the x-axis, and the center of mass for a system with $m_1 = 8$ at $(2, 3)$, $m_2 = 10$ at $(-1, 4)$, $m_3 = 15$ at $(3, -5)$ and $m_4 = 6$ at $(-2, -4)$.

45. Find the center of mass of the lamina of uniform density ρ bounded by the graph of $f(x) = 16 - x^2$ and $y = 0$.

46. Find the center of mass of the lamina of uniform density ρ bounded by the graphs of $y = x^2$ and $y = 4x - 3$.

47. Find the centroid of the region bounded by $f(x) = (x - 1)^2$ and $g(x) = x + 1$.

Check Yourself

1.

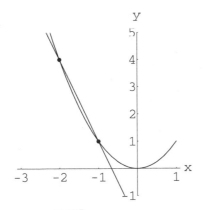

Sketch the graph, or use your graphing calculator

Find the points of intersection:

$$x^2 = -3x - 2 \qquad \text{Set the equations equal}$$

$$x^2 + 3x + 2 = 0 \qquad \text{Add } 3x + 2 \text{ to both sides of the equation}$$

$$(x + 2)(x + 1) = 0 \qquad \text{Factor the left side}$$

$$x = -2, -1 \qquad \text{Set each factor equal to 0 and solve for } x$$

These x-values give the limits of integration.

$$\int_{-2}^{-1} [(-3x-2)-x^2]\,dx \qquad \text{Use (Top Curve – Bottom Curve)}$$

$$= \left(-\frac{3}{2}x^2 - 2x - \frac{x^3}{3}\right)\Bigg|_{-2}^{-1} \qquad \text{Integrate each term}$$

$$= \left(-\frac{3}{2} + 2 + \frac{1}{3}\right) - \left(-6 + 4 + \frac{8}{3}\right) \qquad \text{Substitute } x = -1 \text{ and } x = -2 \text{ into the function and subtract}$$

$$= \frac{1}{6} \quad \textbf{(Area Between Curves)}$$

2. 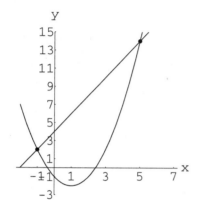 Sketch the graph, or use your graphing calculator

$$x^2 - 2x - 1 = 2x + 4 \qquad \text{Set } f(x) = g(x)$$

$$x^2 - 4x - 5 = 0 \qquad \text{Add } -2x - 4 \text{ to both sides of the equation}$$

$$(x-5)(x+1) = 0 \qquad \text{Factor the left side}$$

$$x = 5, -1 \qquad \text{Set each factor equal to 0 and solve for } x$$

Thus, -1 and 5 are the limits of integration. We will use dx slices, so subtract top curve minus bottom curve:

$$\int_{-1}^{5} [(2x+4) - (x^2 - 2x - 1)]\,dx \qquad \text{Use (Top Curve – Bottom Curve)}$$

$$= \int_{-1}^{5} (4x + 5 - x^2)\,dx \qquad \text{Combine similar terms}$$

$$= \left(2x^2 + 5x - \frac{x^3}{3}\right)\Bigg|_{-1}^{5} \qquad \text{Integrate each term}$$

$$= \left(50 + 25 - \frac{125}{3}\right) - \left(2 - 5 + \frac{1}{3}\right) \qquad \text{Substitute } x = 5, x = -1 \text{ and subtract}$$

$$= 36 \quad \textbf{(Area Between Curves)}$$

3.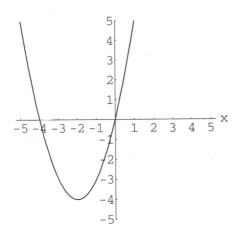

Sketch the curve, or use your graphing calculator

We can see from the sketch that the *x*-intercepts are –4 and 0 (or solve $x^2 + 4x = 0$). We will use dx slices so use top curve ($g(x) = 0$) minus bottom curve ($f(x) = x^2 + 4x$):

$$\int_{-4}^{0} [0 - (x^2 + 4x)]\, dx \qquad \text{Use (Top Curve – Bottom Curve)}$$

$$= \int_{-4}^{0} (-x^2 - 4x)\, dx \qquad \text{Distribute the } -1$$

$$= \left(-\frac{x^3}{3} - 2x^2 \right)\Bigg|_{-4}^{0} \qquad \text{Integrate each term}$$

$$= 0 - \left(\frac{64}{3} - 32 \right) \qquad \text{Substitute } x = 0, x = -4 \text{ and subtract}$$

$$= \frac{32}{3} \quad \textbf{(Area Between Curves)}$$

4.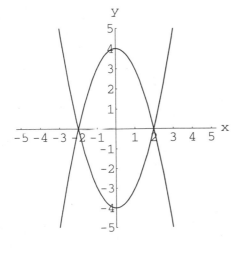

Sketch the curves, or use your graphing calculator

$$-x^2 + 4 = x^2 - 4 \qquad \text{Set } f(x) = g(x)$$

$$-2x^2 + 8 = 0 \qquad \text{Add } -x^2 + 4 \text{ to both sides of the equation}$$

$-2(x^2 - 4) = 0$ Factor

$x = \pm 2$ Add 4 to both sides; take the square root of both sides

$\int_{-2}^{2} [(-x^2 + 4) - (x^2 - 4)]\, dx$ Use (Top Curve – Bottom Curve)

$= \int_{-2}^{2} (-2x^2 + 8)\, dx$ Combine similar terms

$= \left(-\dfrac{2x^3}{3} + 8x \right)\Big|_{-2}^{2}$ Integrate each term

$= \left(-\dfrac{16}{3} + 16 \right) - \left(\dfrac{16}{3} - 16 \right)$ Substitute $x = 2, x = -2$ and subtract

$= \dfrac{64}{3}$

Note that because these curves are symmetric with respect to the y-axis, you could have integrated from $x = 0$ to $x = 2$ and doubled your answer. (**Area Between Curves**)

5.   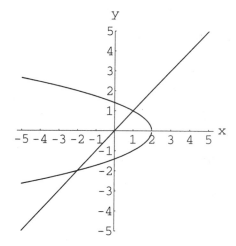   Sketch the curves, or use your graphing calculator

Note that dy slices will work best here. Also note that the right curve is $x = 2 - y^2$ and the left curve is $y = x$. The y values of the intersection will be used as the limits of integration since we are using dy slices.

$y = 2 - y^2$ Set the equations equal to each other

$y^2 + y - 2 = 0$ Add $y^2 - 2$ to both sides of the equation

$(y + 2)(y - 1) = 0$ Factor the left side

$y = -2, 1$ Set each factor equal to 0 and solve for y

$\int_{-2}^{1} [(2 - y^2) - y]\, dy$ Use (Right Curve – Left Curve)

$$= \left(2y - \frac{y^3}{3} - \frac{y^2}{2} \right)\Big|_{-2}^{1} \qquad \text{Integrate each term}$$

$$= \left(2 - \frac{1}{3} - \frac{1}{2} \right) - \left(-4 + \frac{8}{3} - 2 \right) = \frac{9}{2} \quad \textbf{(Area Between Curves)}$$

6.

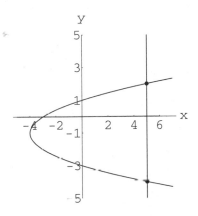

Sketch the curves, or use your graphing calculator

Note: If you want to graph this on a graphing calculator, you may have to solve $x = y^2 + 2y - 3$ for y by completing the square first:

$$x = y^2 + 2y + 1 - 1 - 3 \qquad \text{Add 1 and subtract 1}$$

$$x = (y+1)^2 - 4 \qquad \text{Factor } y^2 + 2y + 1 \text{ as a perfect square trinomial}$$

$$x + 4 = (y+1)^2 \qquad \text{Add 4 to both sides of the equation}$$

$$\pm\sqrt{x+4} = y+1 \qquad \text{Take the square root of both sides of the equation}$$

$$y = -1 \pm \sqrt{x+4} \qquad \text{Subtract 1 from both sides of the equation}$$

Then graph $y = -1 + \sqrt{x+4}$ and $y = -1 - \sqrt{x+4}$ on the same screen.

The intersection points can be found by setting the equations equal since both are equal to x.

$$y^2 + 2y - 3 = 5 \qquad \text{Set the equations equal to each other}$$

$$y^2 + 2y - 8 = 0 \qquad \text{Subtract 5 from both sides of the equation}$$

$$(y+4)(y-2) = 0 \qquad \text{Factor the left side}$$

$$y = -4, 2 \qquad \text{Set each factor equal to 0 and solve for } y$$

$$A = \int_{-4}^{2} [\, 5 - (y^2 + 2y - 3) \,]\, dy \qquad \text{Use (Right Curve - Left Curve)}$$

$$= \int_{-4}^{2} (-y^2 - 2y + 8)\, dy \qquad \text{Combine similar terms}$$

$$= \left(-\frac{y^3}{3} - y^2 + 8y \right)\Bigg|_{-4}^{2} \qquad \text{Integrate each term}$$

$= 36$ **(Area Between Curves)**

7.

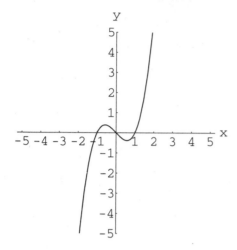

Sketch the graph, or use your graphing calculator

Note that $y = x^3 - x$ is above $y = 0$ on $[-1, 0]$ and below on $[0, 1]$ and so two integrals will be required:

$$\int_{-1}^{0} [\,(x^3 - x) - 0\,]\,dx + \int_{0}^{1} [\,0 - (x^3 - x)\,]\,dx \qquad \text{Use (Top Curve – Bottom Curve) in each interval}$$

$$= \left(\frac{x^4}{4} - \frac{x^2}{2} \right)\Bigg|_{-1}^{0} + \left(-\frac{x^4}{4} + \frac{x^2}{2} \right)\Bigg|_{0}^{1} = \frac{1}{2} \quad \textbf{(Area Between Curves)}$$

8.

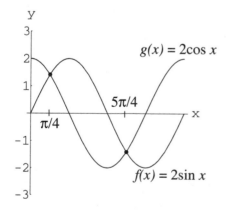

Find the intersection points on $\left[0, \dfrac{5\pi}{4} \right]$:

$2\sin x = 2\cos x$ \qquad Set the functions equal to each other

$\tan x = 1$ \qquad Divide both sides of the equation by $2\cos x$

$x = \dfrac{\pi}{4}, \dfrac{5\pi}{4}$ \qquad These are special angles; tangent is positive in quadrants I and III

Observe your sketch to see that $g(x)$ is the top curve on $\left[0, \dfrac{\pi}{4}\right]$ and $f(x)$ is the top curve on $\left[\dfrac{\pi}{4}, \dfrac{5\pi}{4}\right]$.

Thus two integrals will be required.

$A = \displaystyle\int_0^{\pi/4} (2\cos x - 2\sin x)\,dx + \int_{\pi/4}^{5\pi/4} (2\sin x - 2\cos x)\,dx$ Set up two integrals

$= (2\sin x + 2\cos x)\big|_0^{\pi/4} + (-2\cos x - 2\sin x)\big|_{\pi/4}^{5\pi/4}$ Integrate each term

$= [\,(\sqrt{2} + \sqrt{2}) - (0 + 2)\,] + [\,(\sqrt{2} + \sqrt{2}) - (-\sqrt{2} - \sqrt{2})\,]$ Substitute the upper and lower limits of integration

$= 2\sqrt{2} - 2 + 4\sqrt{2} = 6\sqrt{2} - 2$ Simplify your answer   **(Area Between Curves)**

9.

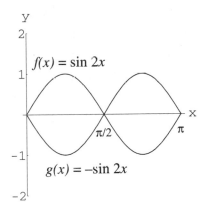

Sketch the curves, or use your graphing calculator

$f(x) = \sin 2x$

$g(x) = -\sin 2x$

Since these curves are reflections across the x-axis, they must intersect at $0, \dfrac{\pi}{2}$, and π. $f(x) = \sin 2x$ is the top curve on $\left[0, \dfrac{\pi}{2}\right]$ and $g(x) = -\sin 2x$ is the top curve on $\left[\dfrac{\pi}{2}, \pi\right]$, so two integrals will be required.

$A = \displaystyle\int_0^{\pi/2} [\,(\sin 2x) - (-\sin 2x)\,]\,dx + \int_{\pi/2}^{\pi} [\,(-\sin 2x) - (\sin 2x)\,]\,dx$ Set up two integrals

$= \displaystyle\int_0^{\pi/2} 2\sin 2x\,dx - \int_{\pi/2}^{\pi} 2\sin 2x\,dx$ Combine similar terms within each integral

$= -\cos 2x\big|_0^{\pi/2} + \cos 2x\big|_{\pi/2}^{\pi}$ Integrate each term

$= 4$ Substitute in the limits of integration and simplify   **(Area Between Curves)**

10.

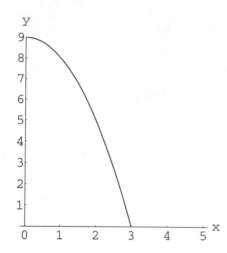

By sketching the plane curve, we can envision the three-dimensional figure that will be formed by revolving the plane curve about the x-axis. We will use dx slices, which form discs, to find the volume.

$$V = \int_0^3 \pi (9 - x^2)^2 \, dx$$ Use the volume formula for discs $V = \int_a^b \pi r^2 dx$, where r is the radius of a disc

$$= \pi \int_0^3 (81 - 18x^2 + x^4) \, dx$$ Factor out the constant π

$$= \pi \left[81x - 6x^3 + \frac{x^5}{5} \right]\Big|_0^3$$ Integrate each term

$$\approx 129.6\pi \approx 407.2$$ Use the limits of integration and your calculator to complete the integration

(The Disc Method)

11.

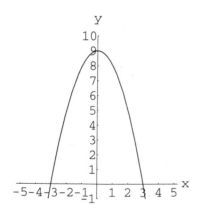

Sketch the graph and note that dy (horizontal) slices work nicely here. The slices begin at $y = 0$, and end at $y = 9$, making the limits of integration 0 and 9. The radius is now x, so we solve for x:

$$y = 9 - x^2$$ Given

$$y - 9 = -x^2$$ Subtract 9 from both sides of the equation

$$\pm\sqrt{-y + 9} = x$$ Multiply both sides of the equation by -1; take the square root of both sides

We use only the positive square root since we are concerned with the side of the parabola where $x \geq 0$.

$$V = \int_0^9 \pi \left(\sqrt{-y+9} \right)^2 dy \qquad \text{Use the formula for the volume using discs}$$

$$= \pi \int_0^9 (-y+9)\, dy \qquad \text{Factor out the constant } \pi$$

$$= \pi \left(-\frac{y^2}{2} + 9y \right) \Bigg|_0^9 \qquad \text{Integrate each term}$$

$$\approx 40.5\pi \approx 127.2 \quad \textbf{(The Disc Method)}$$

12.

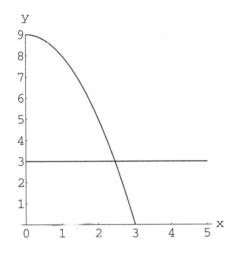

Graph the required region. Note that the slices are dx slices running from $x = 0$ to the intersection of $y = 3$ and $y = 9 - x^2$ or

$$3 = 9 - x^2 \qquad \text{Subtract 9 from both sides of the equation}$$

$$6 = x^2 \qquad \text{Multiply both sides of the equation by } -1$$

$$\sqrt{6} = x \qquad \text{Take the square root of both sides of the equation}$$

The radius is $y - 3$ or $(9 - x^2) - 3 = 6 - x^2$. So,

$$V = \int_0^{\sqrt{6}} \pi \left(6 - x^2 \right)^2 dx \qquad \text{Use the formula for the volume using discs}$$

$$= \pi \int_0^{\sqrt{6}} (36 - 12x^2 + x^4)\, dx \qquad \text{Factor out the constant } \pi; \text{ multiply } \left(6 - x^2 \right)^2$$

$$= \pi \left[36x - 4x^3 + \frac{x^5}{5} \right] \Bigg|_0^{\sqrt{6}} \qquad \text{Integrate each term}$$

$$\approx \pi \left(\frac{96\sqrt{6}}{5} \right) \approx 147.75 \quad \textbf{(The Disc Method)}$$

13. Graph the required region and note that dy slices can be used. The limits of integration are 0 and 8. The radius of the disc is $x-1$ or $\sqrt{-y+9}-1$. So, $V = \int_0^8 \pi\left(\sqrt{-y+9}-1\right)^2 dy$ **(The Disc Method)**

14. $V = \int_0^{\frac{\pi}{2}} \pi\left(\cos x + 1\right)^2 dx$ **(The Disc Method)**

15.

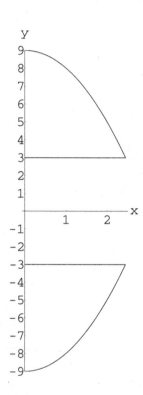

Graph the required region. Note that the dx slice is a washer (a disc with a hole), where the outer radius is y and the inner radius is 3. So,

$$V = \pi\int_0^{\sqrt{6}}\left[\left(9-x^2\right)^2 - 3^2\right]dx \qquad \text{Use } V = \pi\int_a^b \left(\left[R(x)\right]^2 - \left[r(x)\right]^2\right)dx$$

$$= \pi\int_0^{\sqrt{6}}\left(81 - 18x^2 + x^4 - 9\right)dx \qquad \text{Square } \left(9-x^2\right)^2 \text{ and } 3^2$$

$$= \pi\left[72x - 6x^3 + \frac{x^5}{5}\right]\Bigg|_0^{\sqrt{6}} \qquad \text{Combine similar terms; integrate each term}$$

$$\approx \pi\left(105.8\right) \approx 332.4 \quad \text{(The Washer Method)}$$

16.

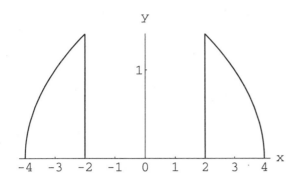

The *dy* slices form washers with outer radius x (or $4 - y^2$) and inner radius 2.

$$V = \pi \int_0^{\sqrt{2}} \left[(4 - y^2)^2 - 2^2 \right] dy \quad \text{Use } V = \pi \int_a^b \left([R(x)]^2 - [r(x)]^2 \right) dx$$

$$= \pi \int_0^{\sqrt{2}} [16 - 8y^2 + y^4 - 4] \, dy \quad \text{Square}$$

$$= \pi \int_0^{\sqrt{2}} (12 - 8y^2 + y^4) \, dy \quad \text{Add similar terms}$$

$$= \pi \left[12y - \frac{8}{3}y^3 + \frac{y^5}{5} \right]\Big|_0^{\sqrt{2}} \quad \text{Integrate each term}$$

$$\approx \pi \left(\frac{112\sqrt{2}}{5} \right) \approx 33.17 \quad \textbf{(The Washer Method)}$$

17.

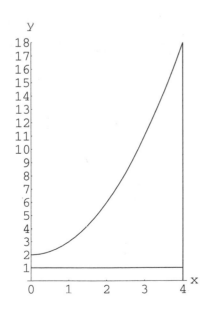

Graph the required region. Note that *dx* slices form washers with outer radius y (or $x^2 + 2$) and inner radius 1.

$$V = \pi \int_0^4 \left[(x^2 + 2)^2 - 1^2 \right] dx \quad \text{Use } V = \pi \int_a^b \left([R(x)]^2 - [r(x)]^2 \right) dx$$

$$= \pi \int_0^4 (x^4 + 4x^2 + 3)\, dx \qquad \text{Square; add similar terms}$$

$$= \pi \left[\frac{x^5}{5} + \frac{4x^3}{3} + 3x \right] \Big|_0^4 \qquad \text{Integrate each term}$$

$$\approx \pi \left(\frac{4532}{15} \right) \approx 949.2 \quad \textbf{(The Washer Method)}$$

18.

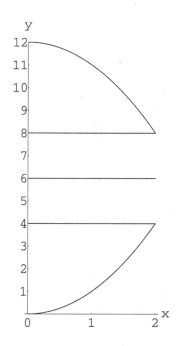

Graph the required region. Note that dx slices form washers with outer radius $6 - y$ (or $6 - x^2$) and inner radius 2 (use $6 - 4 = 2$). From your graph, when $y = 4, x = 2$, so the limits of integration are 0 and 2.

$$V = \pi \int_0^2 \left[(6 - x^2)^2 - 2^2 \right] dx \qquad \text{Use } V = \pi \int_a^b \left([R(x)]^2 - [r(x)]^2 \right) dx$$

$$= \pi \int_0^2 (32 - 12x^2 + x^4)\, dx \qquad \text{Square; add similar terms}$$

$$= \pi \left[32x - 4x^3 + \frac{x^5}{5} \right] \Big|_0^2 \qquad \text{Integrate each term}$$

$$\approx \pi \left(\frac{192}{5} \right) \approx 120.6 \quad \textbf{(The Washer Method)}$$

19. Form dx shells where x is the radius of a shell and $y = x^2 + 4$ is the height of a shell.

$$V = 2\pi \int_0^1 x (x^2 + 4)\, dx \qquad \text{Use the formula for the volume using shells}$$

$$= 2\pi \int_0^1 (x^3 + 4x)\, dx \qquad \text{Multiply}$$

$$= 2\pi \left[\frac{x^4}{4} + 2x^2 \right] \Bigg|_0^1 \qquad \text{Integrate each term}$$

$$= \frac{9}{2}\pi \quad \textbf{(The Shell Method)}$$

20.

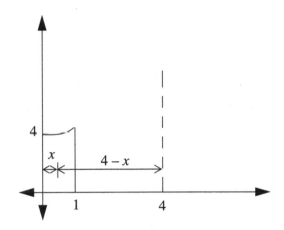

Graph the region. Measure the radius *from* the line $x = 4$. That is, $r = 4 - x$. The height is $y = x^2 + 4$.

$$V = 2\pi \int_0^1 (4-x)(x^2 + 4)\, dx \qquad \text{Use the formula for finding the volume using shells}$$

$$= 2\pi \int_0^1 (4x^2 - x^3 - 4x + 16)\, dx \qquad \text{Multiply}$$

$$= 2\pi \left[\frac{4}{3}x^3 - \frac{x^4}{4} - 2x^2 + 16x \right] \Bigg|_0^1 \qquad \text{Integrate each term}$$

$$= 2\pi \left(\frac{181}{12} \right) = \frac{181\pi}{6} \approx 94.8 \quad \textbf{(The Shell Method)}$$

21. Using the Shell Method with dy slices, radius $= 4 - y$, height $= x \, (= 2y)$, so

$$V = 2\pi \int_0^2 (4-y)\, 2y\, dy \qquad \text{Use the formula for finding the volume using shells}$$

$$= 2\pi \int_0^2 (8y - 2y^2)\, dy \qquad \text{Multiply}$$

$$= 2\pi \left[4y^2 - \frac{2y^3}{3} \right] \Bigg|_0^2 \qquad \text{Integrate each term}$$

$$= 2\pi\left[\left(16 - \frac{16}{3}\right) - 0\right] \qquad \text{Substitute } x = 2, x = 0, \text{ and subtract}$$

$$= \frac{64}{3}\pi \qquad \text{Simplify}$$

(The Shell Method)

22.

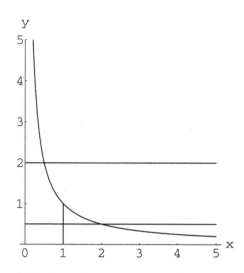

Begin by drawing a sketch similar to the one shown. Envision the small triangular region being revolved around the line $y = 2$ to form a three-dimensional object. If we use shells to find the volume of the object, they will be dy shells. The radius of a typical shell will be measured from the line $y = 2$ and, therefore, the radius of a shell is $2 - y$. The height of a shell will be measured from the curve $y = \frac{1}{x}$ (the right curve) to the curve $x = 1$ (the left curve). Since we are using dy shells, solve $y = \frac{1}{x}$ for x: $x = \frac{1}{y}$. Then:

$$V = 2\pi\int_{1/2}^{1} (2 - y)\left(\frac{1}{y} - 1\right)dy \qquad \text{Use the formula for finding volume using shells} \qquad \textbf{(The Shell Method)}$$

23. a) $V = \pi\int_{0}^{1} [(x - 1)^2]^2 dx$

b) $V = 2\pi\int_{0}^{1} y(-\sqrt{y} + 1)\, dy$. Be careful to use the left branch of the parabola when solving for x:

$$y = (x - 1)^2 \qquad \text{Given}$$

$$-\sqrt{y} = x - 1 \qquad \text{Take the square root of both sides of the equation}$$

$$-\sqrt{y} + 1 = x \qquad \text{Add 1 to both sides of the equation}$$

(The Shell Method and The Disc Method)

24. $f'(x) = \dfrac{3}{2}x^{1/2}$ Find $f'(x)$

   $[f'(x)]^2 = \dfrac{9}{4}x$ Find $[f'(x)]^2$ (Square the first derivative)

   $L = \displaystyle\int_2^4 \sqrt{1 + \dfrac{9}{4}x}\,dx$ Use the formula for arc length

   $\quad = \dfrac{4}{9}\displaystyle\int_2^4 \left(1 + \dfrac{9}{4}x\right)^{1/2}\left(\dfrac{9}{4}\right)dx$ Write the square root as an exponent of 1/2

   $\quad = \dfrac{4}{9}\cdot\dfrac{2}{3}\left(1 + \dfrac{9}{4}x\right)^{3/2}\Big|_2^4$ Integrate using the Power Rule

   $\quad = \dfrac{8}{27}[10^{3/2} - (5.5)^{3/2}] \approx 5.5$ **(Arc Length)**

25. a) $d = \sqrt{(3-0)^2 + (8-0)^2} = \sqrt{73} \approx 8.5$ Find arc length using the distance formula

   b) Find the equation of the line containing $(0,0)$ and $(3,8)$:

   $y - 8 = \dfrac{8}{3}(x-3)$ Use $y - y_1 = m(x - x_1)$

   $y = \dfrac{8}{3}x$ Multiply; add 8 to both sides of the equation

   $y' = \dfrac{8}{3}$ Find the first derivative

   $(y')^2 = \dfrac{64}{9}$ Square the first derivative

   $L = \displaystyle\int_0^3 \sqrt{1 + \dfrac{64}{9}}\,dx$ Use the formula for arc length

   $\quad = \dfrac{\sqrt{73}}{3}\displaystyle\int_0^3 dx$ Factor out $\dfrac{\sqrt{73}}{3}$

   $\quad = \dfrac{\sqrt{73}}{3}x\Big|_0^3$ Integrate

   $\quad = \sqrt{73} \approx 8.5$ **(Arc Length)**

26. $f'(x) = \dfrac{1}{2}x^3 - \dfrac{1}{2x^3}$ Find the first derivative

   $[f'(x)]^2 = \left(\dfrac{1}{2}x^3 - \dfrac{1}{2x^3}\right)^2 = \dfrac{1}{4}x^6 - \dfrac{1}{2} + \dfrac{1}{4x^6}$ Square the first derivative

$$L = \int_4^{16} \sqrt{1 + \frac{1}{4}x^6 - \frac{1}{2} + \frac{1}{4x^6}} \, dx \qquad \text{Use the formula for arc length}$$

$$= \int_4^{16} \sqrt{\frac{1}{4}x^6 + \frac{1}{2} + \frac{1}{4x^6}} \, dx \qquad \text{Simplify the radicand}$$

Now factor the radicand as a perfect square trinomial:

$$= \int_4^{16} \sqrt{\left(\frac{1}{2}x^3 + \frac{1}{2x^3}\right)^2} \, dx \qquad \text{Use } a^2 + 2ab + b^2 = (a+b)^2$$

$$= \int_4^{16} \left(\frac{1}{2}x^3 + \frac{1}{2}x^{-3}\right) dx \qquad \sqrt{a^2} = a, \, a > 0$$

$$= \left(\frac{1}{8}x^4 - \frac{1}{4}x^{-2}\right)\Big|_4^{16} \qquad \text{Integrate each term}$$

$$= \left(8192 - \frac{1}{1024}\right) - \left(32 - \frac{1}{64}\right) \approx 8160.0 \quad \textbf{(Arc Length)}$$

27. $y' = -\sin x \qquad$ Find the first derivative

$[y']^2 = \sin^2 x \qquad$ Square the first derivative

$L = \int_0^{\pi/4} \sqrt{1 + \sin^2 x} \, dx \qquad$ Use the formula for arc length $\qquad$ **(Arc Length)**

28. Since $y = \dfrac{1}{x} = x^{-1}$:

$y' = -x^{-2} \qquad$ Find the first derivative

$[y']^2 = (-x^{-2})^2 = x^{-4} \qquad$ Square the first derivative

$L = \int_1^4 \sqrt{1 + x^{-4}} \, dx \qquad$ Use the formula for arc length $\qquad$ **(Arc Length)**

29. $f'(x) = 2x \qquad$ Find the first derivative

$[f'(x)]^2 = 4x^2 \qquad$ Square the first derivative

$S = 2\pi \int_0^1 x^2 \sqrt{1 + 4x^2} \, dx \qquad$ Use the formula for surface area

Since the problem only requires that an integral be set up, stop here. $\quad$ **(Area of a Surface of Revolution)**

30. $y = \dfrac{1}{2}x^3 \qquad$ Given

$f'(x) = \dfrac{3}{2}x^2 \qquad$ Find the first derivative

$[f'(x)]^2 = \dfrac{9}{4}x^4$ Square the first derivative

$S = 2\pi\displaystyle\int_0^1 \dfrac{1}{2}x^3\sqrt{1+\dfrac{9}{4}x^4}\,dx$ Use the formula for surface area

$= \pi\left(\dfrac{1}{9}\right)\displaystyle\int_0^1 9x^3\sqrt{1+\dfrac{9}{4}x^4}\,dx$ Let $u = 1+\dfrac{9}{4}x^4$ so $du = 9x^3dx$; Multiply by 9 inside and 1/9 outside

$= \dfrac{\pi}{9}\left(\dfrac{2}{3}\right)\left(1+\dfrac{9}{4}x^4\right)^{3/2}\Big|_0^1$ Integrate using $\displaystyle\int u^n du = \dfrac{u^{n+1}}{n+1}$

$= \dfrac{2\pi}{27}\left(\left(\dfrac{13}{4}\right)^{3/2}-1\right) \approx 1.13$ **(Area of a Surface of Revolution)**

31. $x = \dfrac{1}{3}y^3$ Given

$x' = y^2$ Find the first derivative

$(x')^2 = y^4$ Square the first derivative

$S = 2\pi\displaystyle\int_0^1 \dfrac{1}{3}y^3\sqrt{1+y^4}\,dy$ Use the formula for surface area

$= \dfrac{2}{3}\pi\dfrac{1}{4}\displaystyle\int_0^1 4y^3\sqrt{1+y^4}\,dx$ Let $u = 1+y^4$ so $du = 4y^3du$; Multiply inside by 4 and outside by 1/4

$= \dfrac{1}{6}\pi\dfrac{2}{3}(1+y^4)^{3/2}\Big|_0^1$ Integrate using $\displaystyle\int u^n du = \dfrac{u^{n+1}}{n+1}$

$= \dfrac{\pi}{9}(2^{3/2}-1) \approx 0.64$ **(Area of a Surface of Revolution)**

32. $f(x) = 2\sin x$ Given

$f'(x) = 2\cos x$ Find the first derivative

$[f'(x)]^2 = 4\cos^2 x$ Square the first derivative

$S = 2\pi\displaystyle\int_0^{\pi/4} 2\sin x\sqrt{1+4\cos^2 x}\,dx$ Use the formula for surface area

Since the problem only requires that an integral be set up, stop here.   **(Area of a Surface of Revolution)**

33. Since the force is not variable (the weight of the desk doesn't change), use Work = Force × Distance.
     $W = (100 \text{ pounds})(2 \text{ feet}) = 200 \text{ foot-pounds}$ **(Work)**

34. Note that 1.4 kilograms is a mass, not a force. First convert it to a force using the given fact that acceleration
     due to gravity is 9.8 m/s^2. Force = mass × acceleration due to gravity.

$F = (1.4 \text{ kg}) (9.8 \text{m/s}^2) = 13.72$ Newtons. Then

$W = (13.72 \text{ Newtons})(0.53 \text{ meters}) = 7.27$ joules. **(Work)**

35. The work done involves a variable force, so we use $W = \int_a^b F(x)\, dx$. First find k, the spring constant:

$F = kx$ Force is proportional to distance

$10 = k(6)$ Substitute $F = 10$ and $x = 6$

$\dfrac{5}{3} = k$ Divide both sides by 6 to solve for k, the spring constant

$W = \int_0^{10} \dfrac{5}{3} x\, dx$ Use the formula for work with $k = 5/3$

$= \dfrac{5}{6} x^2 \Big|_0^{10} \approx 83.3 \text{ in} \cdot \text{lb}$ **(Work)**

36. First find the distance the spring is stretched when the 30-N force is used:

$x = 50 \text{ cm} - 40 \text{ cm} = 10 \text{ cm} = 0.1 \text{ m}$

$F = kx$ Find the spring constant, k

$30 = k(0.1)$ Substitute $F = 30$ and $x = 0.1$

$k = 300$ Divide both sides by 0.1

$W = \int_{0.5}^{0.6} 300 x\, dx$ Use the formula for work

$= 150 x^2 \Big|_{0.5}^{0.6}$ Integrate

$= 150(0.11) = 16.5 \text{ J}$ **(Work)**

37. Note that the force needed to move a horizontal slice of water is equivalent to the weight of a slice. Using the fact that water weighs approximately 62.4 pounds per cubic foot,

$F = \left(62.4 \dfrac{\text{lb}}{\text{ft}^3}\right) (\text{volume of a slice})$

$= \left(62.4 \dfrac{\text{lb}}{\text{ft}^3}\right) (\pi (10)^2 \Delta y \text{ ft}^3)$ where Δy is the thickness of a slice of water

$= 6240 \pi \Delta y \text{ lb}$

$W = \int_0^{40} 6240 \pi (40 - y)\, dy$ since a slice y ft from the bottom must be moved $(40 - y)$ ft to get it to the top

$= 6240 \pi \left[40 y - \dfrac{y^2}{2}\right]\Big|_0^{40} = 6240 \pi (800) \approx 1.57 \times 10^5 \text{ft} \cdot \text{lb}$ **(Work)**

38. Note that the only change from problem #37 is that the slices of water moved are between 20 feet and 40 feet from the bottom:

$$W = \int_{20}^{40} 6240\pi \, (40 - y) \, dy$$

$$= 6240\pi \left[40y - \frac{y^2}{2} \right] \Bigg|_{20}^{40}$$

$$= 6240\pi \, (800 - 600) \approx 3.9 \times 10^6 \text{ ft} \cdot \text{lb} \quad \textbf{(Work)}$$

39. Note that the volume of a slice is $V = (5 \text{ ft}) \, (6 \text{ ft}) \, (\Delta y \text{ ft}) = 30\Delta y \text{ ft}^3$

$$W = \int_0^4 (62.4) \, (30) \, (4 - y) \, dy$$

$$= 1872 \left[4y - \frac{y^2}{2} \right] \Bigg|_0^4 = 14{,}976 \text{ ft} \cdot \text{lb} \quad \textbf{(Work)}$$

40. Use $F = \delta \int_{y_1}^{y_2} hL \, dy$ where $\delta = 62.4 \text{ lb/ft}^3$ (density of water), $h = -y$ (depth of a slice y feet if we place the axes so that the water level is at $y = 0$ and the bottom of the gate is at $y = -3$ and the top of the gate is at $y = -1$) and $L = 3$ feet since the length of each slice is 3 feet. Then,

$$F = 62.4 \int_{-3}^{-1} (-y) \, (3) \, dy$$

$$- \, 62.4 \left(\frac{3y^2}{2} \right) \Bigg|_3^{-1} \approx 749 \text{ lb.} \quad \textbf{(Fluid Force)}$$

41. Placing the axes as shown, the depth of a horizontal slice is $(-y)$.

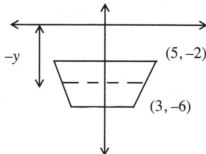

To find the length of a slice, find the equation of the line containing $(5, -2)$ and $(3, -6)$:

$$y - (-2) = \frac{-6 - (-2)}{3 - 5} (x - 5) \qquad \text{Use } y - y_1 = m \, (x - x_1)$$

$$y + 2 = 2 \, (x - 5) \qquad \text{Simplify each side of the equation}$$

We need the x-value, so solve for x:

$$y + 2 = 2x - 10 \qquad \text{Multiply}$$

$\dfrac{y + 12}{2} = x$ Add 10 to both sides of the equation; divide both sides of the equation by 2

Each slice is $2x$ long or $2\left(\dfrac{y + 12}{2}\right) = y + 12$.

$F = 62.4\displaystyle\int_{-6}^{-2} (-y)\,(y + 12)\,dy$ Use the formula for fluid force

$= -62.4\displaystyle\int_{-6}^{-2} (y^2 + 12y)\,dy$ Multiply; factor out -1

$= -62.4\left[\dfrac{y^3}{3} + 6y^2\right]\Big|_{-6}^{-2}$ Integrate each term

$= -62.4\left[\left(-\dfrac{8}{3} + 24\right) - \left(-\dfrac{216}{3} + 216\right)\right] \approx 7654$ lb.   **(Fluid Force)**

42. Place the axes at the center of the window. Then the equation of the circle is $x^2 + y^2 = 1$. Then

   $x = \pm\sqrt{1 - y^2}$ and the length of a horizontal slice is $2\sqrt{1 - y^2}$. The depth of a slice is $10 - y$.

   $F = 62.4\displaystyle\int_{-1}^{1} 2\sqrt{1 - y^2}\,(10 - y)\,dy$. Note that answers may vary based on the initial location of the axes

   (although completing the integration would yield the same number of pounds per pressure).   **(Fluid Force)**

43. Use $\bar{x} = \dfrac{\text{moment}}{\sum m} = \dfrac{10\,(3) + 20\,(5) + 30\,(-6)}{10 + 20 + 30} = -\dfrac{50}{60} = -\dfrac{5}{6}$. This result implies that the system would be

   balanced if the fulcrum were placed at $x = -\dfrac{5}{6}$. **(Moments and Center of Mass)**

44. $M_y = 8(2) + 10(-1) + 15(3) + 6(-2) = 39$

   $M_x = 8(3) + 10(4) + 15(-5) + 6(-4) = -35$

   $m = 8 + 10 + 15 + 6$ Find the total mass

   $(\bar{x}, \bar{y}) = \left(\dfrac{39}{39}, -\dfrac{35}{39}\right) = \left(1, -\dfrac{35}{39}\right)$ Use the formula for center of mass   **(Moments and Center of Mass)**

45. $\bar{x} = 0$ since the graph of $f(x) = 16 - x^2$ is symmetric with respect to the y-axis.

   $\text{mass} = \rho\displaystyle\int_{-4}^{4} (16 - x^2)\,dx$

   $= \rho\left[16x - \dfrac{x^3}{3}\right]\Big|_{-4}^{4} = \dfrac{256}{3}\rho$

   $M_x = \dfrac{\rho}{2}\displaystyle\int_{-4}^{4} (16 - x^2)^2\,dx$

$$= \frac{\rho}{2} \int_{-4}^{4} (256 - 32x^2 + x^4) \, dx \qquad \text{Square } (16 - x^2)^2$$

$$= \frac{\rho}{2} \left[256x - \frac{32x^3}{3} + \frac{x^5}{5} \right] \Bigg|_{-4}^{4} \qquad \text{Integrate each term}$$

$$= \frac{\rho}{2} \left[\left(1024 - \frac{2048}{3} + \frac{1024}{5} \right) - \left(-1024 + \frac{2048}{3} - \frac{1024}{5} \right) \right] \qquad \text{Substitute } x = 4, x = -4 \text{ and subtract}$$

$$= \frac{\rho}{2} \left(\frac{16384}{15} \right) = \frac{8192}{15} \rho \qquad \text{Simplify}$$

Then $\bar{y} = \dfrac{\dfrac{8192}{15}\rho}{\dfrac{256}{3}\rho} = \dfrac{32}{5}$ **(Moments and Center of Mass)**

46. Sketch the region and find the points of intersection of the boundary curves.

$$x^2 = 4x - 3 \qquad \text{Set the equations equal to each other}$$

$$x^2 - 4x + 3 = 0 \qquad \text{Add } -4x + 3 \text{ to both sides of the equation}$$

$$(x - 3)(x - 1) = 0 \qquad \text{Factor the left side}$$

$$x = 3, x = 1 \qquad \text{Set each factor equal to 0 and solve for } x$$

$$m = \rho \int_{1}^{3} [(4x - 3) - x^2] \, dx \qquad \text{Find the total mass}$$

$$= \rho \left[2x^2 - 3x - \frac{x^3}{3} \right] \Bigg|_{1}^{3} = \frac{4}{3} \rho$$

$$M_x = \frac{\rho}{2} \int_{1}^{3} \left[(4x - 3)^2 - (x^2)^2 \right] dx \qquad \text{Find the moment}$$

$$= \frac{\rho}{2} \int_{1}^{3} (16x^2 - 24x + 9 - x^4) \, dx \qquad \text{Simplify the integrand}$$

$$= \frac{\rho}{2} \left[\frac{16}{3}x^3 - 12x^2 + 9x - \frac{x^5}{5} \right] \Bigg|_{1}^{3} \approx 6.13\rho$$

$$M_y = \rho \int_{1}^{3} x \left[(4x - 3) - x^2 \right] dx \qquad \text{Find the moment}$$

$$= \rho \int_{1}^{3} (4x^2 - 3x - x^3) \, dx \qquad \text{Simplify the integrand}$$

$$= \rho \left[\frac{4}{3}x^3 - \frac{3x^2}{2} - \frac{x^4}{4} \right] \Bigg|_{1}^{3} \approx 2.67\rho$$

$$\bar{x} \approx \frac{2.67\rho}{1.33\rho} \approx 2 \qquad \text{Find the } x\text{-coordinate of the center of mass}$$

$$\bar{y} \approx \frac{6.13\rho}{1.33\rho} \approx 4.6 \qquad \text{Find the } y\text{-coordinate of the center of mass} \qquad \textbf{(Moments and Center of Mass)}$$

47. Centroid implies uniform density. The curves intersect at:

$$(x-1)^2 = x+1 \qquad \text{Set the equations equal to each other}$$

$$x^2 - 3x = 0 \qquad \text{Square } (x-1)^2; \text{ add } -x-1 \text{ to both sides of the equation}$$

$$x(x-3) = 0 \qquad \text{Factor the left side}$$

$$x = 0, x = 3 \qquad \text{Set each factor equal to 0 and solve for } x$$

$$m = \int_0^3 [(x+1) - (x-1)^2]\, dx \qquad \text{Find the total mass}$$

$$= \int_0^3 (-x^2 + 3x)\, dx = \frac{9}{2}$$

$$M_x = \frac{1}{2}\int_0^3 [(x+1)^2 - (x-1)^4]\, dx = \frac{36}{5} \qquad \text{Find the moment}$$

$$M_y = \int_0^3 x[(x+1) - (x-1)^2]\, dx \qquad \text{Find the moment}$$

$$= \int_0^3 (-x^3 + 3x^2)\, dx = \frac{27}{4}$$

$$(\bar{x}, \bar{y}) = \left(\frac{\frac{27}{4}}{\frac{9}{2}}, \frac{\frac{36}{5}}{\frac{9}{2}} \right) = \left(\frac{3}{2}, \frac{8}{5} \right) \qquad \text{Find the centroid}$$

(Moments and Center of Mass)

Grade Yourself

Circle the question numbers that you had incorrect. Then indicate the number of questions you missed. If you answered more than three questions incorrectly, you will then have to focus on that topic. If a topic has less than three questions and you had at least one wrong, we suggest you study that topic also. Read your textbook, a review book, or ask your teacher for help.

Subject: Applications of Integration

Topic	Question Numbers	Number Incorrect
Area Between Curves	1, 2, 3, 4, 5, 6, 7, 8, 9	
The Disc Method	10, 11, 12, 13, 14, 23	
The Washer Method	15, 16, 17, 18	
The Shell Method	19, 20, 21, 22, 23	
Arc Length	24, 25, 26, 27, 28	
Area of a Surface of Revolution	29, 30, 31, 32	
Work	33, 34, 35, 36, 37, 38, 39	
Fluid Force	40, 41, 42	
Moments and Center of Mass	43, 44, 45, 46, 47	